GUNS

AN ILLUSTRATED HISTORY
OF ARTILLERY

E. EGG · J. JOBÉ · H. LACHOUQUE · P. E. CLEATOR · D. REICHEL

GUNS

AN ILLUSTRATED HISTORY
OF ARTILLERY

EDITED BY JOSEPH JOBÉ

PATRICK STEPHENS LIMITED
WITH
EDITA · LAUSANNE

ISBN 0 85059 071 X

CONTENTS

PREFACE

Underlying the invention, development and successive uses of guns, the history of artillery reveals also the ever increasing tempo of progress and the course followed by our industrialized civilization.

If the first guns had a range of only a few hundred yards and usually burst their barrels after firing a dozen or so shots, this was due to the fact that the craftsmen of those days were obliged to invent the tools and the end product at the same time. Progress was inevitably slow and groping. Some 300 years had to pass before the seventeenth century dawned, when guns could be produced which ended their life through wear and not untimely explosion.

The first rational system of artillery came into being about 1763. Only then were gun-carriages, spare parts and projectiles made to the same standards as the guns and manufactured with sufficient accuracy to be interchangeable. From 1815 onwards, the modernization of equipment followed a parallel course to the rapid strides being made in scientific knowledge and technical skills. Range was increased by the introduction of rifled barrels, screw or quoin breeches, recoiling gun-carriages and new types of powders and projectiles. Artillery's combat potential was considerably increased, as was its autonomy, by the adoption of recoilless guns and self-propelled gun-carriages. Revolutionary techniques were also developed, not the least momentous being the miniaturization of the atom bomb into an atomic shell.

As for the actual employment of artillery, it could be said that the first generation of guns spread terror, the second razed castles and fortified positions while the third mowed down the close ranks of enemy infantry.

The development of this powerful force, together with the corresponding social changes and architectural trends, led to diversification and reorganization of the troops' duties under such great military leaders as Charles XII of Sweden, Frederick II of Prussia, the Duke of Marlborough and Napoleon I. Artillery was used as a battering ram; it was also an incarnation of the legendary dragon, roaring and belching fire at the will of princes and kings.

Centuries of technical improvements transformed artillery into a decisive arm which had its own part to play in the furtherance of political schemes and military strategy. Cannon boomed out in Africa and Asia, punctuating colonial conquests. From one side of the world to the other, international affairs were resolved by the salvoes of gunboats. A new term crept into the language: "Gunboat diplomacy".

Aircraft and, subsequently, the rocket took over some of the functions once reserved exclusively for the artillery. Now, given even more mobility by the self-propelled gun-carriages, artillery could perform part of the duties traditionally assumed by the cavalry in the past.

Until the eighteenth century, the logistics of artillery regiments remained strictly empirical. Under Napoleon, these questions assumed quite different proportions. Guns and supplies might be needed in Austria or Spain, horses had to be available in Lisbon or Moscow while gunpowder, cannon-balls and transport were required all over Europe. Once, on the actual field of battle, Napoleon ordered 10,000 cannon-balls to be brought into the line within an hour, so that his guns could pound one sector of his enemy's positions. The ability and virtuosity of his lieutenants enjoyed free rein and were responsible for many brilliant victories, such as the Battle of Wagram. The enormous distances to be covered over the immense areas of the theatres of operations, as well as the complexity of the equipment, necessitated the establishment of military and civil services, specialized headquarters' staff and troops. Meteorology, cartography, electronics and management techniques soon became accepted, everyday features of the artillery regiment. Behind each battery, factories, laboratories, intelligence services, research departments and weather forecasting bureaux sprang up and proliferated. During the nineteenth century, the first industrialists modelled their organizational methods on those of the army. Today, the vertical military structure, the chain of command of the past, has given way to a reticular system borrowed from modern management techniques.

A close study of the history of artillery seems to reveal a strange phenomenon. The faster the rate of technical improvements, the slower was acquired the understanding and knowledge of their potentialities. The acceleration of technical progress would appear to be concomitant with a deceleration in the rate of its application, a "constant factor" to compensate for the folly of our day and age.

The authors of this history of artillery have described various aspects, each selecting the one that appeared most interesting or most typical of the period under review. In this way, the most significant events are highlighted and the story enriched by a wealth of varied and often hitherto unpublished illustrations.

J. JOBÉ

FROM THE BEGINNING
TO THE BATTLE
OF MARIGNANO · 1515

The emergence, and subsequent development of artillery from the middle of the fourteenth century onwards was to change the ideas of military commanders and the thinking of the combatants. To fully appreciate its effects and evolution, particularly in the early stages, the technical and psychological aspects should be considered separately.

From the technical point of view, every effort was made from the outset to increase the range and penetration power of the projectiles, to improve accuracy of fire, weapon mobility and to increase the rate of fire. But working methods and the materials used in the Middle Ages were rudimentary, so progress was unsure and very slow. Skills in foundry work were still elementary and the available iron was difficult to work. Crude iron could only be used for small calibre cannon. It was not until the European copper mines in the Tyrol and Saxony started operating during the fifteenth century that bronze was first used in the manufacture of guns. This marked improvement in materials allowed greater accuracy of fire to be obtained. Methods and instruments enabling the target to be sighted appeared only at the end of the fifteenth century. It was even more difficult to accelerate the rate of fire, for lack of charges ready for firing. Cannon were loaded by the muzzle with powder and ball, each time the gun was to be fired. The penetration power of the projectiles depended solely on the force of impact; reliable time-fuzed shells were not introduced until the nineteenth century. Exploding grenades, launched by means of a strip of cord or leather, were thrown by hand, and only used in exceptional circumstances as artillery projectiles. The most marked improvements concerned gun mobility.

However, truly mobile gun-carriages, enabling the artillerymen to keep up with the army, did not appear until the fifteenth century.

The psychological effect of artillery had more immediate and visible results. It transformed the mental attitude of the fighting man, who soon could no longer imagine a battle without the rumble of cannon and the preparation by the artillery. He even refused to go into attack without artillery support. The cannon gave him a feeling of superiority, even if the cannonade produced no other effect than the thunder of explosions. The din, the smoke and the yells of enemy soldiers wounded by stray shot combined to create a characteristic sound and visual condition which plunged the infantry into a state of excitement and aggressiveness. The stone or cast iron shot striking the outer walls of fortifications caused clouds of dust and showers of debris inside, thus making a far greater impression on the defenders than they did on the wall. On the other hand, a few direct shots and the noise of the explosions were often enough to panic horses and cause the helter-skelter flight of their riders, sometimes determining the outcome of the battle.

The moral and psychological effect of artillery was a new factor in the evolution of warfare. The noise of cannon was more awe-inspiring than the drumming of horses' hooves during a cavalry charge, the clash of swords and bucklers or the wild shouts of men. It was to be a long time before artillery attained the accuracy or the efficiency of the traditional siege weapons or before light shells were responsible for noticeable losses in the solid formations of infantry. Nevertheless, the roar of cannon gave the illusion of a supernatural power, the guns appearing as harbingers of hell, come to take possession of the battlefield. Warfare became more terrifying; fighting men felt powerless, at the mercy of luck. For this reason alone, all belligerents lost no time in increasing their artillery, the number of guns being considered of more consequence than their actual efficiency. In spite of the drawbacks of an ever larger and slower artillery train, armies encumbered themselves with more and more cannon.

Over and above the sound effects produced by artillery, an uncanny terror was engendered by superstitions circulated about this new weapon and gunpowder, sometimes known as "Devil's lust". Amidst the roar of explosions the devilish rumour circulated that black magic and witchcraft were at work among the artillery and the men who tended the guns. So well established did this diabolical link become, that a certain churchman put down the marked tendency towards bad language displayed by artillerymen to "their commerce with infernal substances."

* * *

Artillery, as we know it, began with the invention of gunpowder, an explosive mixture of saltpetre, sulphur and charcoal. The projectiles, with their accompanying flames, smoke and noise, no longer depended directly on human strength, on the tension of a cross-bow or the power exerted by a lever. The discovery of gunpowder remains an enigma which may well never be cleared up. Roger Bacon, 1214-1294, the scholarly Franciscan monk who lived in England, disclosed the compound in a work dated 1242 or even earlier, entitled *De secretis operibus artis et naturae*. It is a mixture "which produces a lightning flash and a noise of thunder". About the same time, the famous Albertus Magnus of Cologne also revealed the composition of this mixture of saltpetre, sulphur and charcoal. The role of the legendary monk Berthold Schwarz, who lived in Flanders, still

remains a mystery. An old manuscript describes him as "master of the Greek countries", which gives rise to the supposition that he was in touch with the Byzantines and Arabs who, with the Chinese, were the first people to use gunpowder.

The first gunpowder contrivances used for military ends were vase-shaped receptacles. These employed the explosive force of gunpowder to project a missile roughly in the desired direction. At that time, the only known methods of throwing projectiles over any distance were the long bow and the cross-bow, together with siege weapons such as the mangonel and trebuchet, so it is hardly surprising that the first bodies to be propelled by gunpowder were not cannon balls but arrows and, particularly, incendiary arrows and lumps of rock. The new firearm was first used in the siege of towns and fortresses. In order to launch an attack on Trent in 1278, the town of Verona asked Count Meinrad II of the Tyrol to lend them the machines and specialists capable of throwing "iron and fire".

The oldest known illustrations of these "fire tubes" date from 1326. They appear in two manuscripts written by Walter Milimete in honour of Edward III, King of England, namely *De nobilitatibus, sapientiis prudenciis regum* and *De secretis secretorum*. These drawings show a metal bottle fixed on a plank. A burning arrow projects from its narrow neck while a man inserts a rod of hot iron into the bottle through a touch-hole. Visitors to the military museum in Stockholm can see a bronze "gunpowder bottle", 12 ins long and of a calibre of 1.5 ins, made in the fourteenth century.

Mainly used against castle gates and fortifications, these small "fire boxes" had to be brought very close to their objectives. Although improved siege weapons and devices were available at this time, the thunder of this fire seemed a more terrifying, fearsome and fascinating alternative. A very old document tells us "this weapon is dreaded on account of its noise, unbearable to the ear".

There would have been no artillery as we know it had it not been for the invention of gunpowder. The fourteenth century miniature on the opposite page depicts a pyrotechnist busily preparing the explosive mixture by weighing out the ingredients: saltpetre, sulphur and charcoal.

Alle pulū machstu also feruen ein wenn du pulū wellest machen swetz
so nym Salmē vnd swebel als zu andū pulū vnd nym fawl Allbrem
holtz vnd der dz sol vnd pulū dz vnd tū dz an dez kols stat in dz
pulū so beleibe ez weiz vnd schon Jtem wilan ab dū pulū rot Grūn plaw
gel od swarz habn so feib dz pulū als man salmē ferbet also vor ge
schribn stat vnd wirt dennoch starckes guts pulū nemm dz man damit
grosse vn halbn mūz //

Between 1300 and 1350, portable cannon proliferated and became the fashionable weapons. Gradually, stone, iron or lead balls replaced the incendiary arrows. In 1326, the Signory of Florence instructed Reinaldo de Villamagna to construct a bronze cannon for iron balls. In 1338, a cannon and gunpowder were used against Puy-Guillaume in France while, in 1343, Petrarch alludes to wooden instruments "firing metal nuts, spreading flashes of lightning and peals of thunder". In 1338, the official ledgers of Rouen mention payment for iron tubes, arrows and gunpowder. The following year, the town council of Cambrai ordered five iron and five bronze cannon. In 1341, the town of Lille recorded the use of a "thunder box". In 1340, during the siege of Terni, the Pope's troops used two iron bombards called *Tromba marina* or "marine blast-pumps". In 1327, reference was made to Edward III using "Crakys of war" against the Scotch, and in 1346, at the battle of Crecy, a contempory account describes the English use of bombards "which with fire throw little balls to frighten and destroy horses". Cannon were certainly used by the English in the siege of Calais the following year, but as the daily supply of gunpowder was

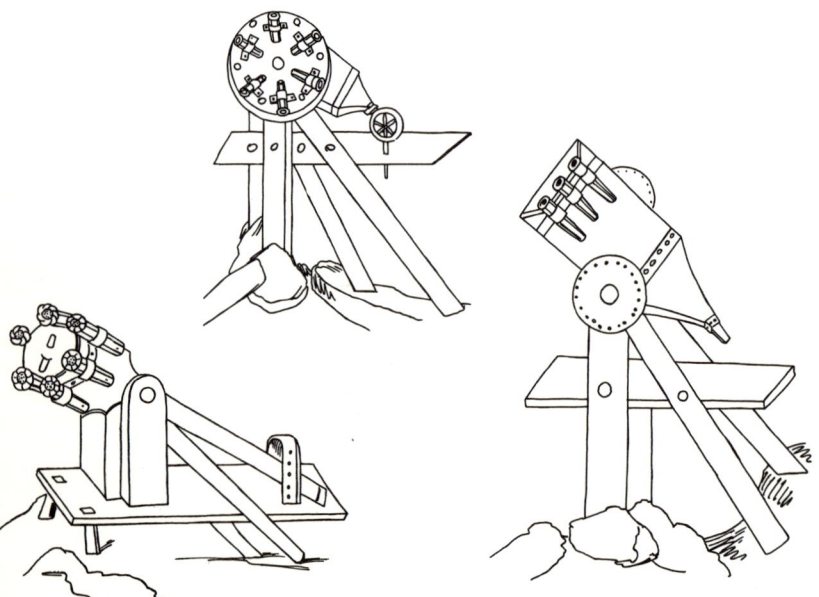

From the appearance of the very first guns, there was a sudden proliferation of different types. Although foundry techniques limited production to small, rudimentary pieces, the novelty of the new arm stimulated many inventive minds whilst arousing the interest of those vying for power.

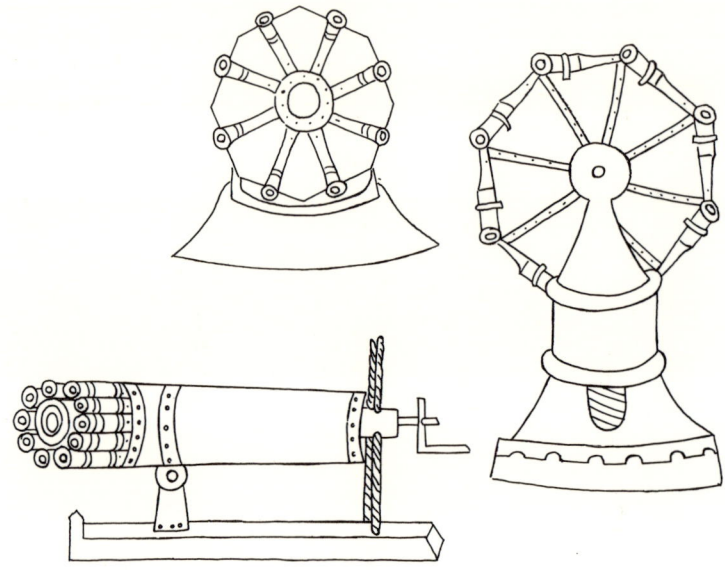

At the beginning of the fifteenth century, artillerymen combined several barrels on the same gun-carriage to intimidate their adversaries. The booming of these primitive contrivances inspired terror and awe, particularly since such a demonstration of evil power was thought to be the result of black magic and witchcraft.

three to four ounces, they made but a small contribution. In 1364, five hundred small cannon, 9 ins. in length and firing lead shot, were made for the town of Perugia. In 1346, the town of Aix-la-Chapelle possessed an "iron box", a *bussa ferrea*.

As time passed, it became the practice to give the new arms definite names such as *Büsse*, *Büchse* or *Donnerbüchse* in Germany, *tromba* or *tronum* in Italy, *canon* in France or *bombarda* in Spain. In the first phase, known as the period of portable cannon, developments in artillery seem to have followed an identical course throughout Europe, the "tubes" being in iron or bronze.

* * *

As the use of cannon became more widespread and armies more accustomed to handling them, war leaders began to look for ways in which to increase their efficiency. Very soon, the first giant cannon appeared. During this period from 1350 to 1450, cannon were made in iron and were distinguished by the size of their powder chamber and chase, that part nearest the muzzle which directs the projectile and determines the accuracy of the

This miniature, dating from the beginning of the fourteenth century, shows the first "fire tubes" to be extremely rudimentary. They looked like metal vases horizontally fixed on a heavy, rigid base. The projectile was still merely a large arrow or spear. Shielding their faces, the gunners stand cautiously to one side, for the powder was set off by inserting the red-hot end of a metal rod into the "touch-hole".

trajectory. As a result of numerous unfortunate experiences, it was established that the amount of gunpowder must be so gauged as to cause the discharge of the shot without the cannon bursting itself. It was at this time that artillery was established as an excellent siege weapon, but was not yet particularly adapted to the field of battle. At this stage of its development, it was suited to static, not to mobile warfare.

The increase in calibres and the lengthening of the barrels required new manufacturing techniques. Iron bars, forged like barrel staves, were fitted together to form a hollow cylinder, strengthened by hoops of iron which were intended to contain the pressure of the fired gunpowder, hence the term "barrel". The small size of the foundries and the insufficient heat available at the time precluded the use of the casting techniques of today. These forged iron cannon were characterized by the length of their powder chamber, twice the diameter of the shot, by the length of their chase, which was one and a half times the diameter of the shot and, particularly, by the fact that the bore of the chase was much larger than that of the powder chamber. However, the length of the barrel remained quite

modest and this gave rise to great dispersion of the projectiles. The stone shot then used was sometimes of a considerable diameter. As far as light cannon were concerned, the powder chamber and the chase were made separately and screwed together before firing commenced. This was the origin of breech-loaded guns, a system confined to light arms, the large guns continuing to be muzzle-loaded.

Among those specializing in forging cannon at this time, the best known was a Belgian, Walter von Arle, a native of Arlon who had settled in Trêves. It was he who forged the large iron cannon, firing stone shot, for the town of Cologne in 1370. In 1378 he produced a similar weapon for Augsburg, one for Passau in 1379, and yet another for Frankfurt in 1377. Jakob von Toran supplied Salzburg with the same type of weapon in 1378, while Ulrich von Eichstätten made one for Augsburg in 1377. German gunsmiths, who were now beginning to establish their European supremacy, settled in Venice around 1376. Hughes-the-German worked under the Italian name of Ugo Tedesco in Sienna in 1407, followed by Giovanni de Alemannia, whose name is mentioned in the town's records in 1441, while a certain Walter de Alemannia set up business in Como about

13

Another old miniature illustrates a step forward as the gun is mobile, mounted on a wheeled carriage by means of two iron hoops. The gun cradle can also be tilted in the vertical plane so as to change the elevation and facilitate ranging the weapon.

Many experiments, often fatal, had to be made before it was learned how to correctly measure the charge so as to propel the shot without bursting the barrel. The two artificers above are seen loading a gun with a pear-shaped funnel which serves as a measure.

1451. Others came to work in England, four such names being noted in the royal budget as early as 1424. Side by side with the Germans, the Lombards also made a considerable contribution in Italy. From 1450, master gunsmiths from Milan, Pavia and Piacenza were at work in Sienna.

Several large cannon of this period have been preserved to this day. One of these, dating from around 1430, and which has a calibre of 35 ins can be seen in the War Museum in Vienna; another, from Burgundy, 8 ft 11 ins in length and with a calibre of 14 ins, is in the Basel Arsenal. At the Artillery Museum in Turin there is a cannon with a calibre of 23 ins; at Ghent, a cannon 16 feet

long with a calibre of 25 ins and known as the *Dulle Griet* dates from 1410. Another famous piece, thought to have been made in Mons in 1461-83 and now in Edinburgh Castle, is the 13 ft long *Mons Meg*, which fired a 330 lb stone shot. An old account mentioned that if loaded with 105 lbs of powder, and set at an angle of 45°, *Mons Meg* could hurl an iron ball 1,408 yds, or a stone ball 2,876 yds. The later period of this type of large cannon for stone shot is represented by the *Faule Magd* or the "Lazy Servant Girl". Forged around 1500, it has a calibre of 20 ins and now rests in Dresden.

It is interesting to note that the development of artillery in Germany started in the Free or Imperial

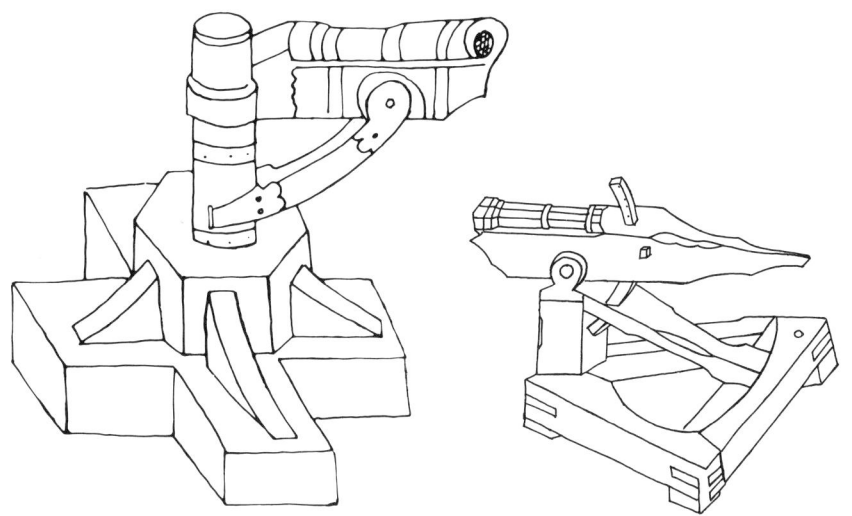

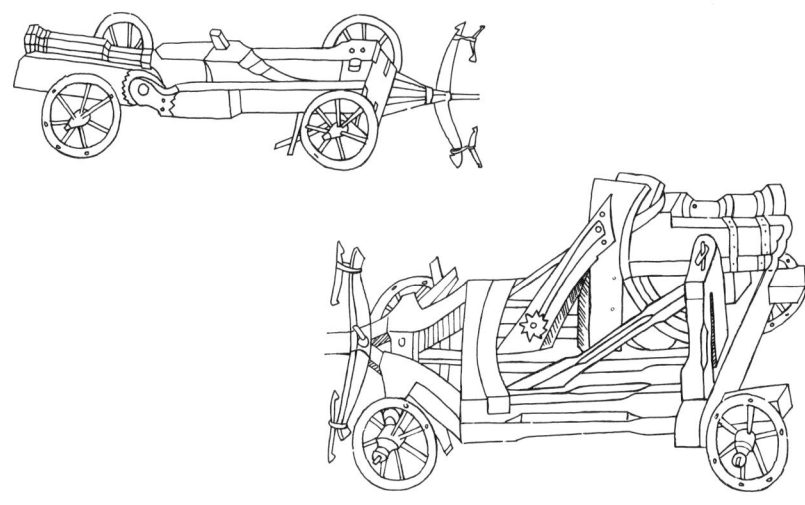

A century after the appearance of the first "fire-boxes", cannon were still quite small and mounted on wooden gun-carriages, as shown in the illustration above.

As time passed and experience was gradually acquired, the addition of wheels to gun-carriages made them more mobile. These were made individually and only a few of each produced.

Cities. They were obliged to defend their own ramparts or to demolish the castles of lesser squires which threatened their lines of communication. In 1391, Frankfurt boasted fourteen large cannon capable of firing 150 to 200 lb stone shot, as well as 101 pieces of lesser calibre.

In 1388, Nuremberg had a large cannon christened *Krimhild* within the town's walls. Twelve horses were needed to pull it and twenty-two more to bring up its equipment and ammunition. Very probably, the *Krimhild* was the legendary "tube for stone balls" whose 1388 construction is attributed to Ulrich Grünewald by military historians. Its cannon balls were reputed to carry a thousand paces and to penetrate walls six ft in thickness. During the same period, Nuremberg also possessed three iron cannon weighing 220 lbs apiece, capable of firing 45-pound shots at the rate of forty a day, as well as two light, wheeled cannon which could fire fifty shots a day. According to contemporary records, when the towns of Mainz and Frankfurt sent their soldiers to lay siege to Tannenberg Castle in 1399, then held by the Count Palatine, the cannon attached to the Frankfurt army breached the wall at the second shot. In 1382, the Archbishop of Salzburg placed a large bombard, firing 100 lb stone shot, at the disposal of Duke Leopold III of Austria, then at war with Francis of Carrara. This cannon's name, the *Trivisana*, has been preserved in history.

Large cannon firing stone projectiles were mounted on horizontal planks or beams. These "stands" were chocked at the back end with planks in order to take the shock of the recoil resulting from discharge of the cannon. In spite of this precaution, however, the gun had to be replaced in position after each shot, and re-aimed at its target.

Master artificers, who were drawn not from the army but from the ranks of craftsmen, wrote instruction manuals to initiate the specialists into the use of artillery, this mysterious and important new arm. The oldest of these manuals is preserved in the Munich State Library, German Codex 600, and dates from around 1350. It is entitled *Die Anleitung Schiesspulver zu bereiten, Büchsen zu laden und zu beschiessen*, "Advice on preparing gunpowder, loading guns and firing".

In 1419, the cruel Hussite wars broke out and created an entirely new situation. In fact, the Hussites had conceived new tactics which crippled the mounted armies of the day. They threw all the burden of the battle on to their infantry and their barricades of charriots. The mobile foot-soldiers thus fought with their backs protected, only drawing away in order to break through or pursue the enemy. Light or medium-weight artillery pieces defended the barricades, from which they fired as from a fortification, but they remained mobile in so far as the army train was mobile. The Hussite artillery consisted of medium cannon mounted on wheels, the *Haufnitzen*, and light cannon fixed on stands, the *Bockbüchse* or *Tarasbüchsen*, which were set up in the waggons, from which firing could be carried on without undue preparation.

About this time, the Germans began to devise a system of names for pieces of artillery firing metal shot. Medium cannon with calibres from 8 ins to 12 ins were called *Haubitzen*, from which the name "howitzer" is derived, and small cannon with calibres from 1 in to 4 ins *Lotbüchsen* or *Tarasbüchsen*. Stone shot brought about the use of larger calibres, as projectiles fired by small-calibre cannon were too light to maintain a flat trajectory.

Certain types of light cannon were breech-loading and had a reserve of two or three charges in the powder chamber, giving them a faster rate of fire. Also, they often fired lead shot with a flat trajectory, comparatively accurately, even at long range. On the other hand, the shorter howitzers, called mortars by the French, carried out high angle (upper register) fire with stone pellets. This division of guns into those capable of firing at angles greater than 45°, the upper register, and those which usually were fired at angles below 45°, the lower register, is maintained today. Upper register guns are known as howitzers and mortars, lower register as cannon or guns. Their specialised uses will become apparant

This design dates from around 1460. The barrel is securely fastened to a heavy carriage with iron brackets. A hand-operated screw enables the gun's angle of fire to be varied.

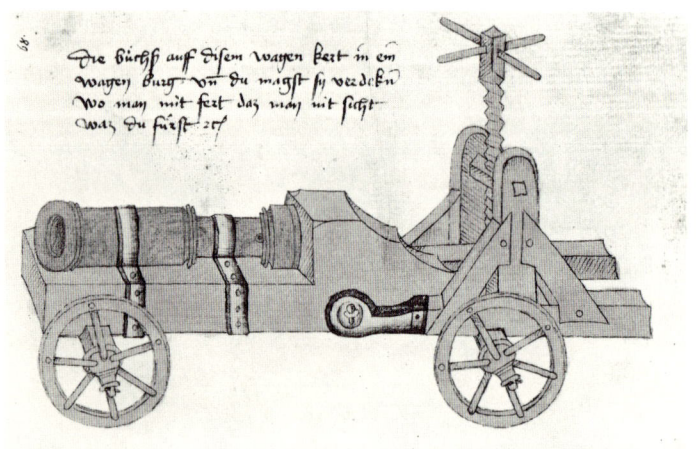

later. A mortar from this period can be seen in the museum at Klagenfurt.

At this stage of development, cannon still lacked the means whereby their firing height could be adjusted to meet the changing situations in combat, for the use of artillery was still primarily confined to siege warfare. For example, in 1422, the Hussites besieged the castle of Karlstein for six months, their forty-one iron cannon firing 1,931 shots. And during the Imperial campaign against the Hussites in 1427, the troops of Nuremberg were armed with a heavy cannon firing 200 lb balls, six small cannon for stone shot and twelve *Tarasbüchsen*.

At the time of the Hussite wars, Conrad Kieser of Eichstätt wrote his *Bellifortis*, a manual for gunners and artificers. His book summarized all that was then known about artillery. Its popularity was so great that it was widely reproduced and plagiarized throughout the fifteenth century. The *Bellifortis*, in common with many other manuals of this kind, suggests combinations of guns formed up in lines or circles, either with a view to increasing the rate of fire or to firing volleys. Valid in theory, this principle was difficult to apply in practice.

From 1450 onwards, bronze began to supersede iron in the manufacture of cannon, except in the case of "chamber cannon", also known as "serpentins" or "serpentines", which were capable of firing at a faster rate because of their reserve charges. Guns made in iron, a material too coarse to ensure regularity in the bore, disappeared around 1520.

Artillery was also making its appearance at sea. From the end of the fourteenth century, certain men-of-war were equipped with iron cannon. As early as 1372, the Spaniards used iron cannon in the battle of La Rochelle. In 1377, thirty-five large ships of the French fleet carried light, breech-loading cannon. These were mounted in a fork on a tripod, and were thus able to be trained horizontally. Superimposed batteries on several decks, which were to revolutionize naval warfare, did not come into general use until the seventeenth century.

* * *

This veritable "engine of war" from the mid-fifteenth century could well have been the forerunner of our self-propelled guns. It combines the three characteristic elements of contemporary artillery: the gun, relative protection for the gun-crew from enemy projectiles and a motive force.

The future now seemed to belong to cannon made in bronze. When cast, this metal made it possible to obtain a constant, uniform thickness which was a great advantage and a marked improvement over wrought iron. Resistance to the explosion of the powder was also uniform. Moreover, bronze was more flexible than iron. The shot fitted the gun calibres better and benefited from the full force released by the explosion of the powder. The loss of dynamic energy was also less, as was the risk of bursting the cannon. Above all, the method of casting ensured a faster and larger production than

had been possible with forging. The difficulty of finding raw material contributed in no small measure to this evolution, for while the tin and copper mines of Flanders, Saxony, the Harz mountains, the Tyrol and Upper Hungary supplied an abundance of material, iron production was slowed down by the impossibility of obtaining sufficient heat.

The Free or Imperial towns, hitherto responsible for the development and progress of artillery, ceded their monopoly to the princes, within whose states lay the valuable tin and copper mines. It was not long before the first bronze cannons appeared, their

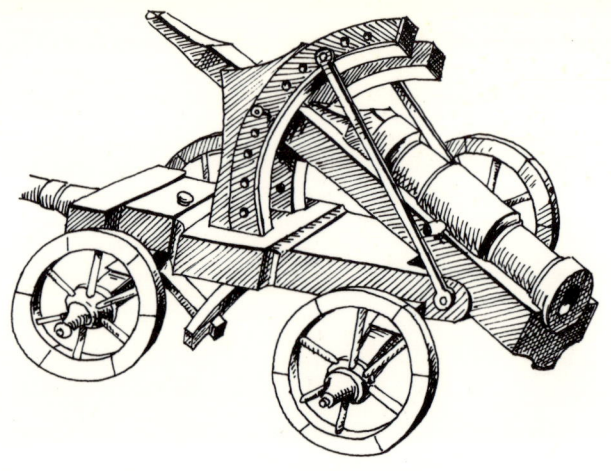

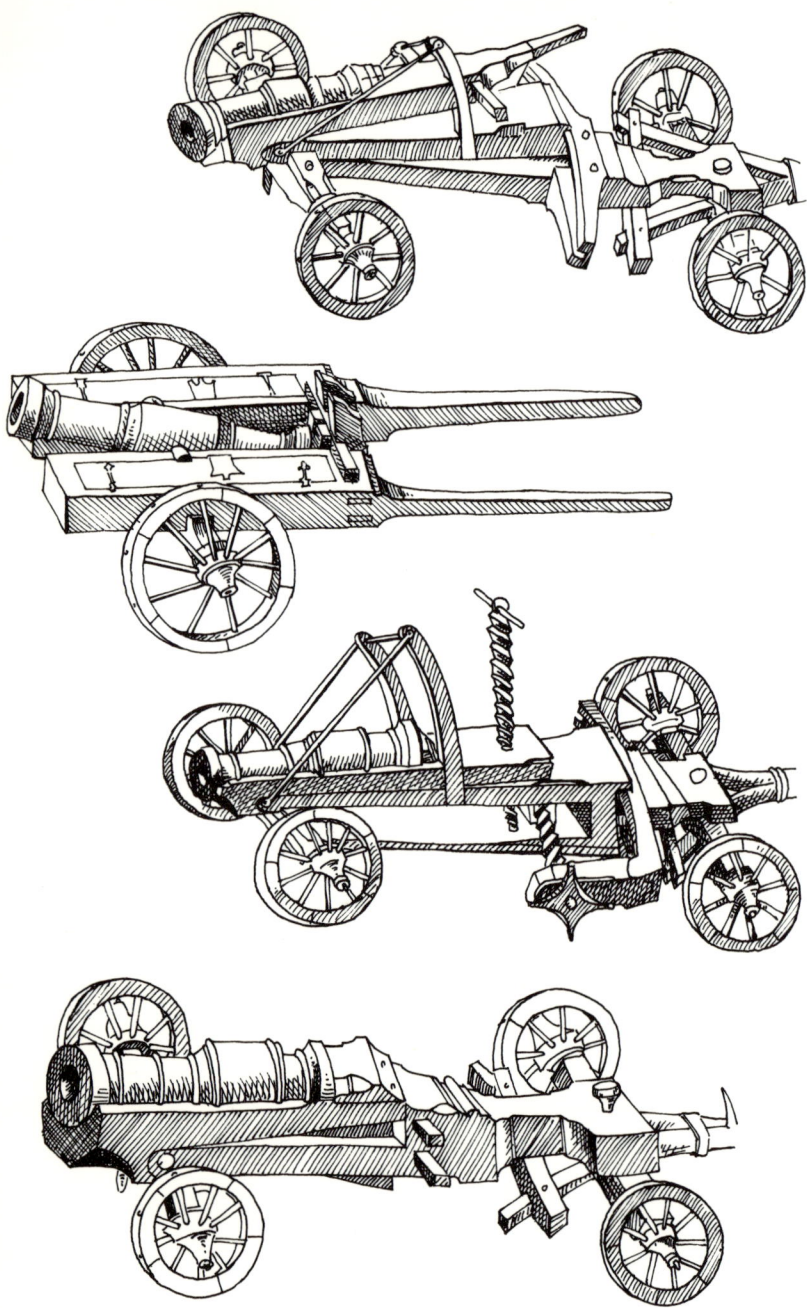

Despite constant improvements, the power and the life of pieces of artillery was still very limited for two main reasons: the weakness of the barrels and the damaging effect of recoil on the gun-carriages.

manufacture following a parallel development to that of iron weapons, at least in the early stages.

In 1346, a quarrel, an arrow or bolt with a square base, and having a calibre of 2 ins and a leaden head, was tested at Tournay. It appears probable that the tin and bronze founders, masters in the art of casting bells and plate, were the first to launch out into the casting of arms. For a long time to come, the same craftsmen were to produce both the cannon of war and the bells of peace, a paradox that has survived even to modern times.

In 1381, the Arsenal of Bologna possessed a copper bombard weighing 361 lbs. In 1370, the Augsburg founders cast twenty bronze cannon with calibres of 11 ins, firing stone balls of about 53 lbs. Even in 1356, the craftsmen of Nuremberg, a trading centre for copper and tin, are known to have constructed bronze cannon for "clients" foreign to the town. The master tin founder, Heinrich Schütz, also cast cannon in 1375. Three years later, records show that Jean d'Aarau cast three cannon for the town of Augsburg, firing balls of 127, 70 and 50 lbs over a distance of a thousand paces. In 1383, Nuremberg delivered guns, which had previously been subjected to test firing, to Duke Stefan of Bavaria. It had already become the custom to carry out "gun trials" to verify the quality of the cast metal and the accuracy of the weapon. A lesser charge than that used when the gun was tested was always used to allow for a known margin of resistance and safety.

The German towns did not have the monopoly of casting and producing guns. Flanders and the Brabant were well-known metallurgical centres. The brass founders of Dinant, heirs to a long tradition which could be traced to Roman times, were the forerunners of the famous foundries of Malines.

In Holland around 1450, which at that time formed part of the lands of the House of Burgundy, an astute inventor perfected the famous "trunnions" which were to bring about important improvements in the development of artillery. Trunnions were short, stubby axles which enabled the cannon to be fixed on to a gun-carriage without hampering their vertical movement. These were cast at the same time as the gun barrel, approximately half-way

An etching by Albrecht Dürer in 1518 depicts a cannon whose barrel has been cast and not forged. Moulded rings round the barrel divide it into several sections, reminiscent of the iron hoops which reinforced forged cannon.

along its length, slightly ahead of the point of balance, so that the weight of the breech would cause the muzzle to rise when pivoted about the trunnions. The invention of the quoin, a wedge-shaped block upon which the breech rested, prior to that of trunnions, already enabled elevation for aim to be adjusted on small field pieces. Another method was to provide uprights either side of the barrel, which were pierced with holes through which iron bars were placed, the barrel resting on them between the uprights. The bars could be moved, and by lowering the rear bar one set of holes, and raising the muzzle set similarly, a greater

elevation could be given. However, these uprights could not stand up for long to the violent impact of recoil. Their use was thus strictly limited to pieces of small calibre. The introduction of trunnions, Flanders' contribution to the development of artillery, led to considerable improvements and their use has survived to this day. The Dukes of Burgundy, Philip the Good and Charles the Bold, whose power depended primarily on their military successes, were the best of clients for the foundries in Flanders. They were the first to engage "masters of artillery", noblemen who had the advantage over master gunners of being familiar with the rudiments of

19

military strategy and tactics. The cannon, hitherto used as an autonomous unit, was no longer an isolated weapon but formed part of an artillery formation. In fact, Charles the Bold was first to introduce a "battery", composed of several guns, to the battle-field. The movement and positioning of the guns was, however, slow and arduous. The battle of Nancy in 1477, during which Charles the Bold was killed, demonstrates this slow speed. The Swiss captured thirty cannon belonging to the Burgundians before they even had time to fire a single shot. A

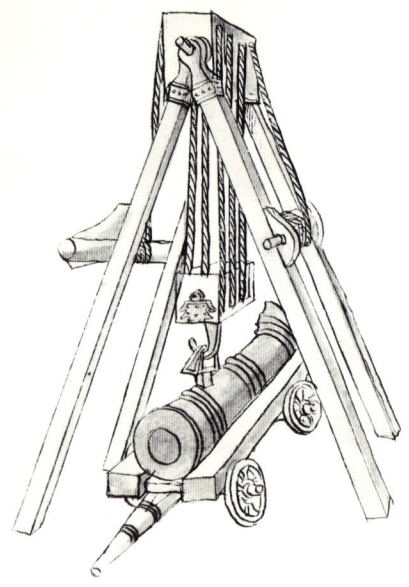

As guns became heavier and heavier, increasing use was made of wind-lasses mounted on sheerlegs in order to lift the barrels.

year earlier, however, at the battle of Morat in 1476, the same antagonists had fired so many stone shots that the ranks of soldiers on both sides had been thrown into confusion. In the course of this battle, the Swiss seized two hundred bombards, mortars and cannon, which clearly shows the weight of artillery used by the Burgundians in their campaigns.

In the battle waged between the Emperor Frederick III and Duke Charles of Burgundy for the town of Neuss in 1474, artillery actually played a decisive role. The Strasburgers, allied to the Emperor, repelled an attack by the Burgundians, crushing them under an hour-long barrage from mortars and bombards. On the following day, the Duke of Burgundy renewed the attack, throwing three corps of infantry, supported by bombards and serpentines, against the Imperial troops, wreaking untold damage, one volley, in particular, killing forty soldiers. Several thousand guilders' worth of gunpowder were used but to no avail; the battle ended in defeat for the Duke.

The Burgundians introduced numerous innovations in the composition and deployment of artillery. They were first in drawing the heavy, thick-set *courtauds*, mounted in wheeled carriages, on to the battle-field. These pieces, also called "quarter cannon" played a decisive role in the development of field artillery. One of these *courtauds*, part of the booty snatched from the Burgundians, can be seen in the Arsenal at Basel. This cannon is 8 ft 4 ins long with a calibre of 9 ins and was produced in the workshops of Jean de Malines in 1474. The princes and townships were quick to recognize the skill of the Burgundian gunsmiths and Johann Stanibruch of Malines was engaged as bombardier by the Republic of Sienna in 1483.

Hans Poppenruyter, who worked in Malines until 1534, was the principal supplier to Henry VIII of England. He cast 144 guns for him, including the famous *Twelve Apostles* which performed such wonders at the battle of Guinegatte in 1479, although in 1513 St John fell from grace by becoming stuck in the mud and then captured.

Therefore, historical records certainly suggest that the Hussites and the Burgundians were responsible for introducing artillery on to the field of battle during the fifteenth century. From this time on-wards, the use of various ancillary equipment, such as the range-card and the quadrant, came into general use. An Italian, Nicholas Tartaglia, wrote a book in 1537-43, which he dedicated to Henry VIII, and which was a compendium of much useful information for gunners. He devised gunner's quadrants, ballistics diagrams, showing "the way of the pellet", and spoke of night fighting, and the best ways of aiming. The range-card, whose arc was divided into twelve degrees, enabled adjustment of the guns' traverse to be made in a regular manner. The metal quadrant served to determine the trajectory and the angle of fire. At around this time, too, fine gunpowder was replaced by granules, increasing its explosive power and diminishing the

From 1490 onwards, the Austrian Emperor Maximilian I concentrated the production of his cannon in the Tyrol which, at that time, boasted the famous foundries of Hans Seelos (1514), Jörg Endorfer (1508) and Peter Löffler (1530). In 1512, Maximilian had 38 double cannon, 24 basilisks, 57 cannon, 137 serpentines and 115 falconets. The arsenal at Innsbruch was the largest in his Empire, being capable of arming, equipping and supplying 10,000 men. This painting by Jörg Kölderer, of about 1507, shows the inner courtyard of the arsenal.

21

risk of accidental firing. From the middle of the fifteenth century, cannon were made longer; it was soon realized that the longer barrel improved the trajectory of the shot and gave greater accuracy.

Siege artillery also made great strides under the influence of the Turks. They had been prompted to construct large cannon for the siege of Constantinople in 1453. The Sultan Mahomet II had thirteen giant cannon cast and they fired four thousand times between 1 April and 10 May, making large breaches in walls twelve yds high and five yds thick. In order to range on the vessels blockading the port, he ordered large mortars to be cast. Their plunging fire annihilated the Byzantine fleet. From then on, European armies adopted the mortar for short range plunging fire. In 1456, the Turks laid siege to Belgrade with 12 giant cannon, 4 mortars and 500 medium calibre guns. Although they fired more than 1,400 rounds daily, they never succeeded in taking the town.

The Turks also secured the services of German gunners, such as Jörg von Meissen, who was taken prisoner by the Knights of Malta during a Turkish attack on Rhodes in 1480. He was later hung for his crimes. Because the use of artillery gained the Turks so many victories, their Balkan enemies lost no time in following their example. Master founder Jean de Zagreb cast two cannon for the burghers of Sienna in 1472.

From the middle of the fifteenth century, artillery-parks were enlarged and diversified. Of the pieces surviving, there is the cannon called *La Bussona*, dating from 1482, belonging to the Knights of St John at Rhodes, which fired 55 lb shot, and a Byzantine cannon dating from somewhere between 1440 and 1456, firing 58 lb rounds. Both of these can still be seen in the Nuremberg museum. At Woolwich, there is a cannon which belonged to the Sultan Mahomet II in 1453, known as "Mahomet's Cannon", which was capable of firing cannon balls weighing over 600 lbs each. It required 30 wagons to move it, 200 men to help 60 oxen draw it, while 250 men went ahead to level the route and strengthen bridges. It took 2 months to travel 150 miles, and then only fired 7 times daily.

Light as it was, the artillery of King Charles VIII of France proved highly effective on the Italian battle-fields during the campaigns of 1494 and 1495. Already, these guns were firing shot made from metal. The "blitzkreig" waged by Charles VIII fired the imaginations of the day and chroniclers were tempted to attribute quasi-supernatural powers to his artillery, thus allaying the bitter humiliation of the defeat suffered by the Renaissance princes. In reality, however, the strength of the French artillery lay less in the quality of its guns than in its excellent organization. The centralization of the administration of the Kingdom of France had enabled the organization of a "weapons system", equipped with an excellent artillery park and also a remount service which the Italian princes could never match. Field artillery can hardly be improvised with bombards, laboriously taken down from their ramparts and harnessed to dozens of yokes of oxen.

The wars waged in Italy by the French compelled the Italians to alter their traditional planning. Until then, the Italian towns and republics had based their strategy on recruiting small armies of mercenaries for relatively short periods. They laid siege to one another, or clashed in small scale engagements. Consequently, their artillery consisted almost entirely of siege guns and wall pieces. It should be added that, at this time, the Italians were not over impressed by fire-arms. The army of Charles VIII was something quite new to them. Its mobility, enabling it to overrun vast areas, reaching as far as Naples, took the towns and princes unawares. The French King's army included 6,000 Swiss mercenaries with their long, dreaded pikes, 400 harquebusiers, 140 mobile cannon, comprising 12 large cannon firing 50 lb metal rounds and 128 light field guns called falcons. All this weaponry added up to a fire-power the like of which the Italians had never imagined possible. Nobody could hold out against the combined effect of artillery and portable fire-arms, the "hedgehog" formed by the spiky rows of halberdiers in the centre and the armoured cavalry on the flanks.

The French had long experience in the making of small-arms, such as culverins and bombards, which complemented the artillery. As early as 1459, a French gunsmith called Master Simon, who manufactured bombards, was registered in Sienna.

Although its results were short-lived, the Italian campaign conducted by Charles VIII nevertheless heralded a new era. German artillery was undeniably the best to be found in Europe but as the Germanic territories were split up into innumerable petty princedoms and fiefs it was not used to the best account. From 1485, however, a large artillery-park began to flourish in the Tyrol. This was mainly due to the wealth of the local copper mines and the work of the Innsbruck founders, Hans Seelos and Jörg Endorfer. In 1486, the Innsbruck arsenal housed eight large cannon, two large mortars with calibres of 24 ins, four medium mortars, five medium cannon, thirteen serpentines mounted as light field guns and fifteen *courtauds* used as heavy field guns, forerunners of the later heavy field artillery. But the Tyrolese artillery was not entirely concentrated in Innsbruck. Reference has been made to three large pieces, thirteen *courtauds*, twenty-seven medium cannon, fifty-four serpentines and thirty culverins outside the town's walls. The Tyrolese artillery, which seems to have undergone a complete transformation between 1480 and 1487, received its baptism of fire during the war against Venice. It played an important part in the victory of Calliano and especially in the capture of the fortress of Rovereto. The Venetians, for their part, possessed excellent artillery, as is shown in a votive picture offered to the parish church of Wilten, near Innsbruck, by the knight Ludwig Klingkhamer, badly wounded by a shot from a Venetian falconet. This seems to be the earliest known pictorial representation of a gun wound.

The *Musée de l'Armée* in Paris houses one of these Tyrolese pieces of ordnance, the *Katharina*. It is 12 feet long with a calibre of 15 ins and was cast in 1487. It is interesting to note, in this connection, that the custom of christening guns goes back to the beginning of the fifteenth century. They were given names which recalled either the merits of their owner, as the *Katharina*, their own virtues or failings, as "The Lazy Girl" or the strength of an animal, such as the *Löwe* or "Lioness". Guns were attributed with a personality, a special character and sometimes mysterious powers.

Not only were guns personified, the custom gradually emerged to embellish and decorate them. Shaped rings formed intersections along the barrel, reminiscent of the iron hoops which were an essential reinforcement of the old forged cannon. Names and armorial bearings perpetuated the identity of the prince who ordered them and the founder who cast them. Illustrative ornamentation evoked the name of the cannon or its type while animals' heads adorned the supports of the barrels. This preoccupation with decoration is undoubtedly typical of the period and the practice was common to all craftsmen. The craftsmen in the foundries lavished particular care on this aspect of their work as if they wished to stress the special, almost loving relationship that existed between them and their creations. The inscriptions often issued a warning or a challenge to the enemy as well. The *Katharina* bears this verse:

> *Die Kateri heiss ich*
> *Vor meiner Gewalt hüt dich*
> *Dans Unrecht straf ich*

In English it reads:

> My name is Catherine.
> Beware of my strength.
> I punish injustice.

The principal states which made up the old German Empire boasted large artillery-parks as early as 1480.

In 1474, the Margrave Albert Achille of Brandenburg, an "experienced Master of Artillery", demanded for his 30,000 troops thirty serpentines, ten mortars and seventy *ribaudequins*, of which twenty should be capable of firing case-shot. This type of cannon was intended to discharge lead pellets at the opposing infantry. In 1486, the Tyrol had about sixty such pieces of its own.

The Bavarian dukes were also able to mount a considerable artillery force. Between 1488 and 1489, the Duchy of Lower Bavaria-Landshut could muster four large cannon, three mortars, nine medium calibre cannon, thirty-one culverins, one

Maximilian

Die vnbekannten sein mir disen tag kurtz
bekanntlichkeit vnns vermag. konig
maximilian vnns beschiiff. oder vns
wirt gen manger milder zueff

Notpuchsen Schiessen eysen

An inventory of Maximilian's artillery was drawn up under his orders. This sumptuous, parchment Zeugbuch, which catalogued every weapon, has been preserved to this day, affording posterity a picture of artillery that was not only powerful but decorated with true Imperial magnificence, as can be seen on the facing page.

On the right, a page from the Zeugbuch prepared for Maxmilian I provides interesting details of the construction of gun-carriages of this period. The gun cradle can be seen as carved out of a single piece of timber. The painter of this miniature has not omitted to depict the carpenter's tools and other equipment such as the broad axe, plumb-line, set-square, nails and trestles.

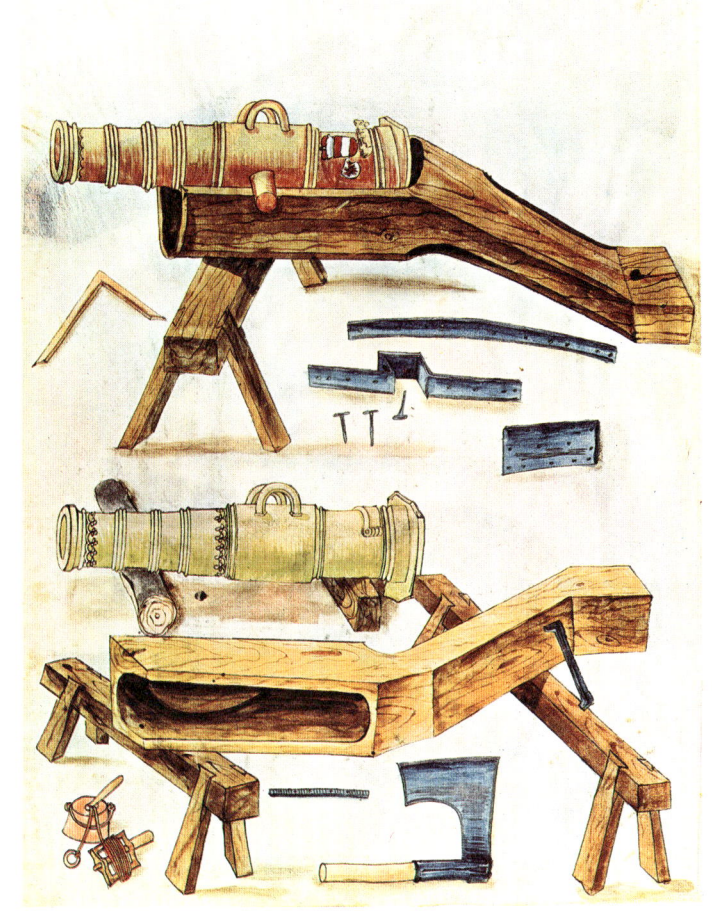

Maxmilian's artillery was made up of heavy cannon, also known as "wall-breakers", and light field guns or culverins, as shown below on the left and right. The gun carriage of the light pieces was lengthened by shafts for horses; some were even fitted with boxes containing tools or spare parts, including horse shoes.

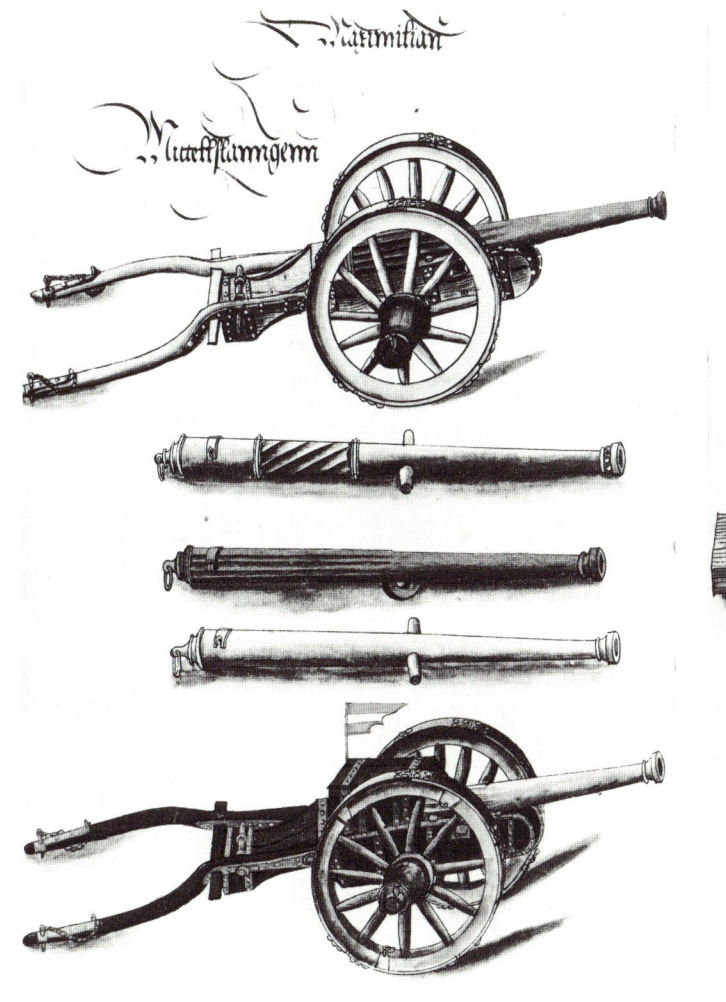

courtaud, seventeen serpentines and eighteen small bombards. The Palatine Counts of Bavaria were proud of the distinguished Master Gunners serving them, who compiled various works based on their experience. Martin Merz was the author of *The Art of Firing Cannon*, published between 1471 and 1475; Philippe Mönch was responsible for *The Book on Warfare and Cannon* in 1496; Louis d'Eybe wrote *The Master Gunner's Manual* in 1500. Martin Merz claims to have personally fired some 372 tons of gunpowder during 1470 and 1471 and to have brought about the destruction of eight castles. He had the following proud inscription engraved on his tomb: "More famous in the art of mathematics and gunnery than anyone".

The last two decades of the fifteenth century were marked by the outbreak of large-scale military campaigns which put an end to the relative balance of power and the ever-precarious peace of the Middle Ages. The great powers joined in battle, an attempt to secure supremacy in Europe. The Spanish, the French, the House of Habsburg, the Turks and the English were the main protagonists in these wars which were characterized by constantly shifting alliances. Italy, Burgundy and Hungary formed the chief battle-fields. These vast and often distant campaigns waged by large armies were to change the concept of military strategy and tactics. The backbone of these armies was formed by the *lansquenets*, as the German foot-soldiers were known, and Swiss, German or Dutch mercenaries. Armed with long pikes, they were drawn up in closely packed formations, bristling with halberds. Harquebusiers, equipped with small fire-arms able to be carried or mounted on tripods, marched in front and all around them. Massed on both flanks, the cavalry held itself in readiness to charge.

Little by little, artillery was to complement and transform this basic concept. Isolated pieces were no longer brought into play but were massed together to form a corps. Despite all the inherent weaknesses due to its slowness, mobile artillery became an important feature in the field. It should be remembered, however, that large guns often fired no more than one shot in the course of a whole battle. The primary objective was to increase artillery's capacity to carry out its main task, the siege and capture of fortresses and strategic towns. These had to be neutralized at all costs, otherwise the entire plan of a campaign had to be modified. The psychological effect of artillery was sometimes quite marked, judging by a Helvetic chronicle referring to the capture of Homburg, near Solothurn; the fortress surrendered after the first round from a serpentine. The defenders of Küssenberg Castle did not even wait long enough for the first shot to be fired. As soon as they saw the enemy positioning its siege guns, they handed over the castle without further ado.

Prominent throughout these wars waged for European supremacy, the Habsburg Emperor Maximilian I, despite the limited resources at his disposal, did not wish to relinquish his lead. Since acquiring the Duchy of Burgundy in 1477, he had been at war with France. From 1495 onwards, he fought the French in Italy. Later, he busied himself defending the West against the invasion of the Turks. He realized that only a powerful army could enable him to make his voice heard among those of the other great powers of that age. So he lost no time in reinforcing his corps of lansquenets, modelled on the regiments of Swiss mercenaries, with the most modern artillery of his time. Having acquired Burgundy and the Tyrol, he availed himself of the skill found in the native foundries.

In Italy, he had had the opportunity of assessing the results of the French and Spanish tactics, as they were already masters in the art of using artillery in the field. During his campaign against the Venetians, he had observed the efforts of the Republic of Venice, the only Italian state to be equipped with artillery, in order to use and adapt his newly acquired knowledge about Turkish artillery to his own needs. With all this information, Maximilian was able to weigh up all the advantages and shortcomings of artillery which, it must be borne in mind, had developed at a lightning but chaotic pace during the latter half of the fifteenth century.

The main weakness of artillery lay in the great diversity of types and calibres, making it impossible to have a rational system of ammunition supply.

Almost every piece required its own size of shot. As a result, the increase in the number of guns on the battle-field did not lead to a proportionate increase in efficiency. After conquering the Tyrol in 1490, Maximilian concentrated his production of fire-arms in Innsbruck. The mountainous features of this area and the fiery nature of its inhabitants seemed to offer possibilities for effective protection and defence. Innsbruck was the home of the large foundries for cannon established by Hans Seelos, who died in 1514, Jörg Endorfer, died 1508, and Peter Löffler whose death is recorded in 1530. Here, also, were the rolling and beating shops where the helmets and cuirasses of the infantry were forged. Pikes, halberds and swords were manufactured at Hall near Innsbruck while iron cannon balls were produced in the town of Absam. The mines at Schwaz supplied both the copper for arming the troops and the silver for paying them. As a gun manufacturer and minter, Maximilian can be considered as the father of the first ordnance supplies establishment, in the modern meaning of the term.

From 1500, the arsenal at Innsbruck was also the central depot for supplies and other equipment. Ten thousand men could be mobilized and fully equipped at once for a large-scale campaign. Secondary arsenals were established in various regions for more precise or limited objectives. Lindau was the base to be used against the Swiss; Brissach against the French; Sigmundskron, near Bolzano, Trent, Goeritz and Verona against the Venetians and French; Osterwitz, Graz and Vienna against the Turks. Maximilian also worked out long-term plans for the development of his artillery. His first "Instruction" in 1500 laid down a programme for two large cannon firing iron shot, eight double cannon firing stone shot, fourteen cannon and numerous 14-pounder field culverins. These were followed in 1562 by three double cannon, twenty-six cannon, fourteen culverins and thirty-two field culverin. The second, dated 1567, stipulated five large cannon, four double cannon, one heavy cannon, six culverins and four field culverins.

The war against the Venetians, lasting from 1508 to 1516, forced Maximilian to build up his artillery strength. In 1512, the Emperor's nine arsenals housed thirty-eight double cannon, twenty-four basilisks, fifty-seven cannon, 137 serpentines and 115 falconets, which comprised the reserve of mobile artillery. With the continuation of the war it became necessary to establish foundries at Rovereto, Trent and Verona. Pierre le Bourguignon, a native of Picardy who died in 1528, and the Master Founder Hans Schnee, also died in 1517, benefited from the lessons of the war then in process to cast improved or more up-to-date pieces, such as falcons, falconets and basilisks in these foundries. The large siege guns were abandoned about 1510 because of their excessive weight, lack of mobility and slow rate of fire. It was impossible to mount them on wheeled carriages and they had to be touched off level with the ground.

So it was that the new artillery came into being, under the direct influence of Maximilian and through the propagation of his ideas. It was divided into two branches: siege artillery and field artillery.

Siege artillery, sometimes called "wall-breakers", comprised four main types of weapons:

the large 48-pounder cannon, abandoned from 1510 onwards.
the demi or double cannon firing rounds of 36 to 50 lbs.
the quarter cannon firing shots of 24 to 40 lbs.
the basilisk, or long culverin, firing rounds of 12 to 24 lbs.

The large cannon making up the heaviest siege artillery were named after legendary women such as Helen, Semiramis and Dido. The Emperor actually entrusted this task to the celebrated classical scholar, Conrad Peutinger. The siege guns most frequently used were the demi-cannon and the quarter cannon, called *Nachtigal* or "nightingale" in the case of the long gun and *Singerin* or "singer" for the shorter gun. The basilisk with its very long barrel was admirably suited to the siege of fortresses because of its accuracy, and the force with which its shot breached the stoutest walls. Basilisks were given the names of animals, such as *Steinbock* or "ibex", *Eidechse* or "lizard", *Krokodil* or "crocodile", or other names evoking their capacities, such as "wall-breaker" or "beater". The demi and the quarter cannon were also suitable for open warfare.

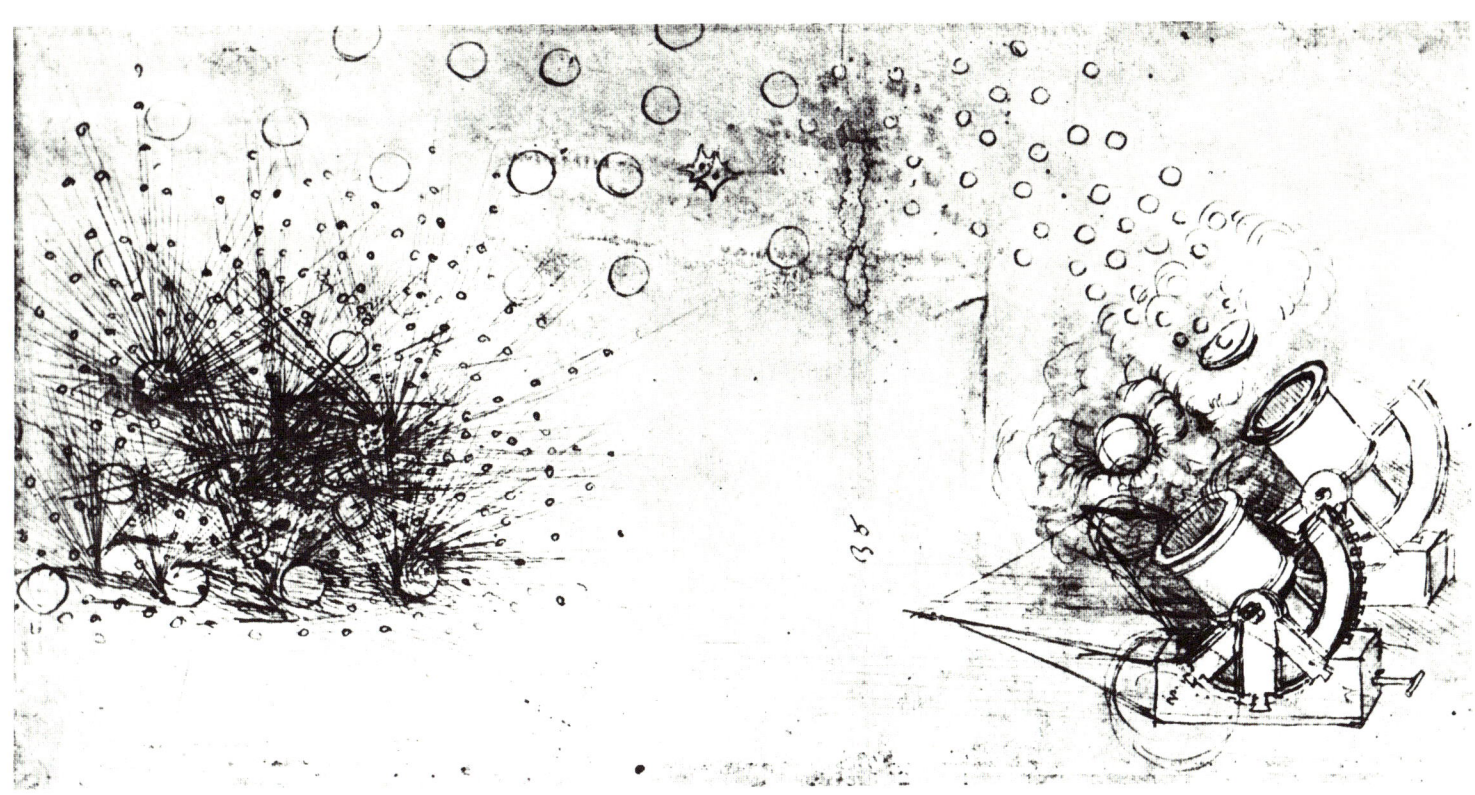

"I also have a type of bombard that is extremely easy and convenient to transport and with which it is intended to shower a veritable hailstorm of small stones and whose smoke will strike the enemy with terror, causing great damage and confusion." This excerpt is from the draft of a letter written by Leonardo da Vinci to Ludovic Sforza about 1482.

Field artillery proper also comprised four types of ordnance:

    the large culverin firing shots of 11 to 16 lbs.
    the culverin firing shots of 6 to 8 lbs.
    the falcon firing 4 lb rounds.
    the falconet with 2 lb shot.

From 1515 onwards, all these guns fired iron shot, with the exception of falconets which were loaded with lead. Field artillery also often fired case-shot, scraps of lead and metal which, on scattering, inflicted heavier damage on the enemy infantry. The heaviest pieces were used against entrenchments, dug-outs and natural or man-made defences such as trenches, ditches, parapets, stockades and other types of barricades. Maximilian ordered iron shot for general use throughout the entire artillery and had his guns mounted on wheeled carriages. Double cannons were pulled by sixteen horses, "singers" were drawn by twelve and large culverins by eight.

All guns of the same type had the same calibre. To ensure the balls were of uniform diameter, metal rings or "templates" were used to check them. At long last, these gauges enabled the standardized, mass production of munitions. On leaving the foundry, every gun-barrel underwent a boring process to standardize the calibre and put a polish on the inside.

In addition to these eight main types of weapons, a series of special pieces existed, such as siege mortars. Large mortars, with calibres of up to 30 ins, fired stone shot while medium and small mortars could also fire incendiary rounds. Light field artillery was

On 2 March 1476, the now Confederated Swiss army defeated the troops of Charles the Bold at Grandson. (See left.) Part of the very powerful artillery belonging to the Duke of Burgundy fell into the victors' hands before there was even time for it to be engaged in battle.

29

reinforced by forged iron cannon, some with interchangeable powder chambers allowing a faster rate of fire. Maximilian concentrated the entire production of these iron cannon in the workshops of Sebald Pögl at Thörl, near Aflenz in Styria. Between 1498 and 1506 this large works supplied 385 field culverins, 619 chamber culverins, 291 small culverins and blunderbusses or *Dorndrel*, the latter resembling the old howitzers and being capable of upper register fire, similar to the mortars.

Small culverins were very often massed in a parallel formation of between six and forty guns, all firing simultaneously. These "organs" inflicted terrible punishment on enemy infantry because of their wide dispersion. Their charge was all the more deadly as it usually consisted of small metal pellets and scraps of lead, similar to the shrapnel shell of a later period. Maximilian also ordered the firing of balls with iron rods from howitzers which, he thought, would produce the same effect.

Between 1500 and 1510, Maximilian ordered twelve large siege cannon to be made. These were of a type similar to those used much earlier, giant pieces loaded level with the ground and firing stone or iron rounds against fortresses and fortifications. They were the last cannon of this type, at least in the Imperial artillery. Their shots had considerable penetration force but, having no accuracy whatever and being extremely difficult to manœuvre, Maximilian finally found it best to abandon them. The Emperor wanted all his guns to be mobile and in constant readiness for firing. Consequently, they had to be fitted with wheeled carriages, as no simple axle would have been strong enough to support the weight of such large siege guns.

Under the reign of Maximilian, decoration on pieces of ordnance reached the height of its glory. The large pieces provided sufficient space to feature all the Imperial coats-of-arms, excellent propaganda for the Empire. The mould in which the *Lauerpfeif* or "Watchman" was cast in 1507 can be seen in the Armorial Museum in Vienna. Maximilian had this gun engraved with the crests of the seven kingdoms he had always claimed as his heritage, namely England, Hungary, Bohemia, Dalmatia, Austria,

Burgundy and Rome, whose imperial eagle is supported by griffins. Rich ornamentation of bands, mouldings and rosettes surrounded these armorial bearings, completed by the emblem of the Golden Fleece and the inscription:

Ich sehe und lauer wie de Hagel und Schauer.
Und heiss darum die Lauerpfeif, nimm hinweg, was ich ergreif.

or, in English,

Like the hail and the storm,
I carry away all I can seize.

All the other large guns owned by Maximilian were also remarkable for their ornamentation and their exotic or highly evocative names such as "The Bewitched Girl", "Madame Humserine", "The Leopard of Wilten" or "The Abominable Leo".

Medium-sized guns and other pieces were decorated much more simply, merely bearing Maximilian's royal arms and the emblem of the Golden Fleece. Basilisks, however, often had their names engraved on the muzzle, accompanied by a dragon's head, as can be seen on the only basilisk that has survived to this day, now in the Basel Arsenal.

The decoration of arms cast after 1510 is even more restrained. This usually comprised a few ornamental details around the muzzle and the base of the breech, and armorial bearings or a shield in the middle of the barrel. Fantasy was giving way to technical progress. More geometric external workmanship complemented the smooth bore. At the same time, tactics on the battle-field were more scientifically applied. Maximilian's cannon, cast around 1515 and now in the German National Museum of Nuremberg, displays a purity of line heralding the Renaissance style.

The immense headway made in the development of artillery under Maximilian was not entirely due to the casting of the new guns; the gunners' training also had a considerable effect. The Master Gunners or Armourers of Innsbruck were the best organizers of artillery at that time. Bartolomé Freisleben, from 1493 to 1509, and Michel Ott d'Achterdingen, from 1515 onwards, were both responsible for the Imperial military supplies and made a vital contribution to the development of artillery. Freisleben, a locksmith by trade, was appointed Master Gunner while Michel Ott, a man of noble descent and excellent education, carried out the duties of "Master of Artillery" as understood by the Burgundians. He was responsible for the maintenance and reinforcement of the artillery-park and the army supplies and also attended to the training of the gunners, Gun Captains and Master Gunners.

As long as the number of guns of all types remained relatively small, the craftsmen who had forged or cast them were used as gun crews. When the artillery-park was enlarged, however, problems of personnel were soon apparent. It was no longer possible to send skilled founders on to the battle-field to serve as artillerymen without jeopardizing production. To replace them in their capacity of Gun Captains or Master Gunners, recruiters enlisted locksmiths, blacksmiths and carpenters, whose training in their trades had given them a basic knowledge which could be applied to loading and unloading the guns from the waggons, manoeuvring, handling and firing them. Maximilian even requisitioned trained painters and masons for these duties. Gradually, their functions became more integrated with the military organization. Gunners, in common with lansquenets, were enlisted for the duration of a campaign and discharged at the end of hostilities.

Maximilian I was thus responsible for many changes and improvements. Artillery owes to him the classification of various types of guns into categories according to their calibres; rationalization in the use of iron shot; the organization of artillery into an independent branch of the army; the promotion of gunners' training and the mounting of cannon on wheeled carriages. Admittedly, all these innovations or improvements were matched in other countries where some were, even earlier, current practice. Maxmilian's merit lay in applying them all simultaneously and methodically in order to make artillery a truly efficient weapon. In his work entitled *La Pirotechnica*, dating from 1450, Vanoccio Biringuccio of Sienna, one of the leading artillery experts of the sixteenth century, notes that "in Maximilian's foundries and arsenals, the art of artillery was the most flourishing in all Christendom". Biringuccio was well qualified to make this judge-

ment. In his youth, he had seen Maximilian's artillery in action on the Italian battle-fields.

The wars in Europe which followed one on the heels of another from 1500 until nearly the end of the century, as well as the example set by Maximilian, led to an increase in the size of artillery-parks and improvements in guns practically everywhere. In Germany, Nuremberg remained an important centre of production and export as it had been in the past. In 1513, Marten Harder cast four large siege guns there for Henry VIII of England.

Henry VIII was another artillery-minded monarch. He appointed a Master Gunner and twelve paid assistants, the first permanent force of gunners in England. When required for action, guns were combined to form a "trayne", served by master gunners, mates, and mattrosses, or assistants. The guns of the "trayne" were called "pieces of Ordnance" (hence the word "Piece" as applied to the gun today) and were named individually and according to size after all kinds of monsters.

The following table gives names, diameter of the bore and weight of the shot of typical pieces. The diameter of the shot was $\frac{1}{4}$ in less than the bore in each case:

Double Cannon (eldest and biggest sorte)	8 $\frac{1}{4}$ in	70-pounder
Double Cannon (ordinary) . .	8 in	64-pounder
Demy Cannon (eldest and biggest sorte)	6 $\frac{1}{2}$ in	38-pounder
Demy Cannon (ordinary) . .	6 $\frac{1}{4}$ in	33-pounder
Culverings (eldest and biggest sorte)	5 $\frac{1}{2}$ in	20-pounder
Culverings (ordinary)	5 $\frac{1}{4}$ in	17-pounder
Demy Culverings (eldest and biggest sorte)	4 $\frac{3}{4}$ in	12 $\frac{1}{2}$-pounder
Demy Culverings (ordinary) .	4 $\frac{1}{2}$ in	10-pounder
Saker (eldest and biggest sorte)	4 in	7 $\frac{1}{4}$-pounder
Saker (ordinary)	3 $\frac{3}{4}$ in	6-pounder
Minyon	3 $\frac{1}{4}$ in	3 $\frac{3}{4}$-pounder
Faucon	2 $\frac{3}{4}$ in	2 $\frac{1}{8}$-pounder
Fauconet	2 $\frac{1}{4}$ in	1 $\frac{1}{8}$-pounder

In addition to these guns, mortars were used extensively by Henry's artillery. In action they were employed to drop large shells over the walls of a besieged city into the town. Henry VIII obtained two gunmakers from the Continent, Peter Bawd and Peter Van Collen, to make large mortars and shells for him in 1543. Some of these mortars measured 11 in and 19 in in diameter, and the shells were stuffed with "wild fire or firewoorkes and a match (fuze) that the firewoorke might be set on fire for to breake in smal peeces, whereof the smallest peece hitting any man would kill or spoile him". Whether they were successful or not is difficult to say, but by 1588 explosive shells were used to good effect. The method of igniting the fuze was either by placing the shell in the bore, fuze towards the charge so that on being fired it would ignite the fuze: or by placing the fuze towards the muzzle, when it was lighted by a match thrust down the bore. It needs no imagination to appreciate the high mortality among those gunners who had to perform this latter task.

In Saxony, where the tin and copper mines had begun to operate, the Hilger family of founders launched into the manufacture of heavy pieces of ordnance, the basis of the Saxon heavy artillery about 1515. A new production centre sprang up in Strasburg, where the famous master Jörg de Guntheim cast cannon for the Emperor, the town of Basel and for the Kings of Spain and England. These were of the same type as had been adopted by the Imperial artillery. Jörg de Guntheim contributed to the replacement of forged cannon-balls by those in cast iron. One of his most noteworthy creations is the remarkable basilisk "The Dragon" which was cast in 1514 and can now be seen in the Arsenal Museum at Basel. This is undoubtedly one of the finest examples of fire-arms surviving from the Maximilian epoch. His compatriot, Matthieu of Strasburg, was killed in 1506 while acting as an armourer in Pisa.

Following the large-scale Italian campaign, France also made great headway in the development of her artillery. Like Maximilian, the French concentrated their efforts on the standardization of calibres and achieving greater gun mobility. Spain which, from 1492, could consider itself an important European power, paid particular attention to the development of light field artillery and small-arms. The Spanish harquebusiers had already proved themselves during the Italian war.

The Republic of Venice, which had borne the brunt of the Italian war from 1508 to 1516, was the the only Italian state capable of vying with the great kingdoms. Venice was the fief of the Alberghetti family who had introduced several distinguished experts in foundry work into the republic, among them the famous Sigismond who lived from 1487 to 1530. It was the Alberghetti workshops that produced the *Colubrinetta* falconet belonging to the ducal palace, one of the oldest cannon in cast iron. It was beautifully embellished by the application of bronze decorations. A certain Giovanni della Tolle, born in 1470 and who died in 1540, was the most celebrated Italian foundry expert. A native of Arbe, he settled in Dubrovnik, and by 1506 had completely transformed the artillery of this town.

The inventory of the artillery owned by the Duchy of Ferrara, which was ruled by the Este family, is known to have been as follows in 1505:

The *Gran Diavolo* or "Great Devil" firing stone rounds of 125 lbs.
7 cannons firing stone shot of 25 lbs.
17 falconets firing iron shot of 4 to 5 lbs.
2 falcons firing iron rounds of 6 lbs.
3 culverins firing iron shot of 25 lbs.
2 small culverins firing iron shot of 10 lbs.
1 cannon firing iron shot of 25 lbs.

Despite its unpretentiousness, the artillery-park owned by the House of Este was noteworthy for its quality. The Ferrara guns were commanded by two dazzling artillerists, Duke Alphonse and his brother, Cardinal Hippolyte d'Este. They had had the political acumen to ally themselves with Maxmilian but, in the field, they fought a series of engagements in which they introduced bold changes.

In the Polesella affair, the brothers' artillery, cleverly concealed in pits dug into the banks of the River Po, "tinged the waters of the river with blood", according to the chronicler Giovio. Although well equipped, the Venetians suffered severe losses and, badly mauled, were obliged to respect the independence of Ferrara for the time being.

Unfortunately, very little is known about the introduction of such concealed firing, by which

As shown to the right of the illustration, the French army, entrenched at Marignano, was attacked by the Swiss infantry on 13-14 September 1515. The French artillery, comprising about 374 guns, inflicted heavy casualties in the Swiss ranks on the first day and, the following day, decided the outcome of the battle by pouring a heavy fire on those Swiss who had managed to face the French lines.

Vauban later achieved such successful results. However, it is known that the Este artillery was placed at the disposal of Gaston de Foix in 1512 and was the deciding factor at the battle of Ravenna. It was quickly brought to bear on the flank of Colonna's Spanish army and cut it to pieces. Gaston de Foix was therefore justified in having waited four days for the artillery trains to arrive before giving battle. The experience of Ravenna enabled the French to win the battle of Marignano, and Machiavelli to establish a veritable doctrine on the active deployment of artillery. Taking these facts into account, it will readily be appreciated how important a part the House of Este played in the development of this arm, more by the brilliance of its Master Gunners than by the size of its park.

Elsewhere in Italy, artillery development was patchy, to say the least. The military importance of Florence, which had boasted three Master Founders, Francesco Telli, Lorenzo Cavallero and Bonacorso di Vittoria in 1496, had dwindled. The town of Milan, particularly exposed and unable to acquire sufficiently strong artillery, was the chief victim of the war waged by the great powers. Nevertheless, Leonardo da Vinci devoted several of his written works to the art of artillery. Studying the relation between the length, calibre and range of cannon, he defined the basic laws of ballistics, drew up plans for several cannon and even suggested replacing the explosive force of gunpowder by the expansion of steam pressure.

Several records from the Maximilian period have been preserved, disclosing the organization of artillery at that time. For instance, in 1504, Duke Albert of Bavaria drew up the following list of guns required for a "small campaign".

33

3 siege cannon firing stone shot of 70 lbs, as well as 200 shot and 60 quintals of gunpowder for each piece.

4 cannon firing shot of 40 lbs, with 250 shot and 50 quintals of gunpowder for each piece.

4 "singers" firing shot of 20 lbs, each supplied with 300 shot and 45 quintals of powder.

6 culverins, firing shot of 11 lbs, each supplied with 300 shot and 24 quintals of powder.

6 demi culverins, firing shot of 8 lbs and each supplied with 350 shot and 18 quintals of powder.

6 falconets firing shot of 6 lbs and each supplied with 400 shot and 12 quintals of powder.

In 1514, the armament recommended for a "major campaign" consisted of:

3 siege cannon with 400 shot.
4 long cannon or "nightingales" with 600 shot.
5 short cannon or "singers" with 800 shot.
6 medium cannon or basilisks with 1400 shot.
8 demi-cannon or falconets.

The complement of personnel corresponding to this inventory of supplies consisted of one "Master-of-Arms", twenty-four Gun Captains or Master Gunners for the heavy guns, fourteen Master Gunners for the small and medium arms, plus sixty-five gunners and auxiliaries. Some 160 horses were required to move this artillery.

Apart from the French campaign in Italy from 1494 to 1495, artillery began to play a truly decisive role during the siege of the Kufstein fortress by Maximilian in 1504. The Emperor personally conducted the siege and commanded seven large cannon firing stone shot, four medium cannon and basilisks, six long and short cannon and ten serpentines or culverins. The castle was destroyed after a cannonade lasting fourteen days. This event inspired a popular song of the day. For the first time, the men of the artillery were featured in the repertory of story-tellers and campaigners. The ballad sung by the defenders of Kufstein had a couplet,

"A great number of guns were seen on the plain
And all began firing at the castle."

The first large concentration of artillery recorded in military history has been traced back to the Italian war of 1508 to 1516, which was characterized by endlessly shifting alliances. Most the great powers of the age were pitted against each other in what

can be undoubtedly be considered as the first "European" conflict. The tactics and strategies employed during this war were extremely varied. The French were supported by an infantry composed of Swiss mercenaries and a considerable artillery-park. Nevertheless, they placed the brunt of the battle on their cavalry which, although heavily armoured, were able to carry out lightning charges. The Spanish relied on their foot-soldiers and light artillery, but even more on their harquebusiers, or troops supplied with small arms. Maximilian trusted in his lansquenets, supported by strong artillery. Venice depended on the fortifications of its strong-holds, for its infantry and light cavalry were insufficiently trained and not powerful enough to intervene successfully in the field.

In 1509, Maximilian attempted to force the well-fortified town of Padua to surrender. He prepared his attack with heavy and prolonged battering from siege cannon and 106 guns on wheeled carriages. A contemporary account relates that "... the gun-fire was so dense on all sides that the air was disturbed over an incredible distance and the ground shook, so that the walls of Padua collapsed in many a place and the towers and houses crashed down in great numbers..." The cannonade lasted from 15 September to 29 September.

A first attack, launched after five days of gun-fire, was broken up by mines detonated by the defenders. This use of mines was an innovation which rapidly came into general use in all sieges.

A lull followed the abortive attack of 20 September. But it was not long before the bombardment was resumed and the cannon fired 1,500 balls against the town in three days, in preparation for a fresh attack on 29 September which met with total defeat. Maximilian raised the siege. Impressive as it was, his artillery-park was nevertheless no match for a citadel as strong as that of Padua.

The Italian campaign was in fact the testing ground for new tactics which consisted of combining the use of different arms, and which had already been successfully used by the French King Charles VIII during his campaign two decades earlier, in 1494. This time, the method was perfected and tested on all the Italian battle-fields where English, French, Spanish, Venetian, Habsburg and Papal troops fought, sometimes on one side, sometimes on the other, according to the shifting coalitions, but all benefiting from the experience.

Henceforth, the brunt of the battle was to rest on the footsoldiers, massed in serried ranks and bristling with pikes, mobile and formidable. All countries depended on the "hedgehog" of their infantry. These troops consisted of Swiss mercenaries, German lansquenets, Spanish pike-bearers and halberdiers. The French possessed armoured cavalry, made up of the King's noble vassals.

In the course of the campaign, artillery proved its worth. It was now a mobile arm and could follow the army from one siege to another, being particularly useful in the defence of beleaguered towns and fortresses. It intervened effectively on the battle-field, spreading confusion and disorder among the enemy ranks poised to attack. Artillery fire broke up the compact formations of infantry with their bristling pikes. However, if the first volley did not produce the desired effect, if the ranks of the enemy infantry remained unbroken and still held battle order, the intervention of artillery was to no avail. Nevertheless, by its presence alone, it played an important role. From now on the infantry were to be given a new mission, to capture the enemy artillery by surprise. As infantry gradually took precedence over cavalry in the field, all the artillery was given greater mobility, resulting in the creation of true field artillery and a rapid increase in the number of guns. Armies abandoned the heavy siege cannon and stone-throwing cannon as too cumbersome to keep up with troop movements.

During the battle of Ravenna between the French and the Spanish in 1512, the Spaniards were the first ever to engage their light artillery in a flanking attack, thus abandoning the position then reserved for the guns and gunners, namely in front of the infantry. This battle was one of the last in which the French cavalry played a decisive role. The Spanish, however, were able to beat an orderly retreat, owing to the protection of their artillery. The engagement had begun with an artillery duel,

the like of which had never been seen on a battle-field before. The Spanish guns killed more than 1,000 German lansquenets from the French army, while one cannon ball alone was said to have mown down 40 men.

In the battle of Novara, which took place from 3 June to 5 June 1513, the Swiss fought the French for the first time. The French infantry was recruited from German mercenaries, enlisted to replace these same Swiss soldiers. Although the French artillery broke down the walls of Novara, the Swiss withstood all attacks and held the town until 8,000 reinforcements arrived and swooped down on the French entrenched camp. Caught under a hail of cannon-balls, they succeeded, nevertheless, in taking possession of twenty-two French cannon, killing all the gunners and losing only 400 men in the effort. They celebrated this doughty deed as their greatest victory, proud of having captured all the enemy artillery while not possessing even one gun themselves. This was the greatest triumph achieved in combat with infantry alone, for the Swiss possessed no cavalry and did not hold artillery in high esteem.

But the Swiss triumph only lasted two years. In 1515, the French resumed hostilities in Italy. Faithful to its traditional tactics, the French army lined up its infantry of German lansquenets, its armoured cavalry composed of the flower of the nobility, and an artillery-park of seventy-two guns, adding up to a total of 55,000 men. The Swiss, who had just lost what artillery they had during the retreat from Milan, were to encounter the French at Marignano on 13-14 September. The French army were entrenched in a camp, protected by stockades and ditches, with its guns deployed. The Swiss launched a frontal attack under sustained fire from the French artillery and suffered heavy losses. They advanced as far as the middle of the French camp where they were obliged to dig in for the night, not having succeeded in completely breaking through the enemy lines. The French artillery continued firing, despite the darkness and the loss of twelve cannon. They still held sixty guns, which blasted away with renewed vigour throughout the following day. The Swiss were unable to approach them.

A diarist reported that "... the French brought all their guns into action. The sky burst forth as if all the furies of the earth and heavens were preparing to swoop down on us...".

To the ceaseless thunder from the cannons, 6,000 French harquebusiers added their crackling discharges. Caught in this hellish fire, the Swiss infantry began to fall back and, for the first time in its glorious history, broke into a rout which the French cavalry turned into a disaster. The troops of the Swiss Confederation left 7,000 dead on the battle-field but, nevertheless, did succeed in evacuating their wounded and the cannon captured from the enemy.

The German lansquenets, drawn up in narrower ranks, had a greater proportion of marksmen than the Swiss, and the fire from their harquebuses swept the compact mass of Swiss troops. The French victory at Marignano was in fact the victory of combined weaponry, the carefully considered application of the military tactics victoriously employed in 1494 by King Charles VIII.

Following these large-scale battles in the field, the Italian war took a new turn, unexpectedly reverting to the siege and attack of fortified positions. In 1516, the French and the Venetians, allies at this time, laid siege to Verona, the gateway to Germany, defended by the Emperor and the Spanish. Jörg von Fruntsperg and Colonna, beset by an adversary equipped with powerful artillery, revealed themselves as masters in the art of defence. A Spaniard succeeded in infiltrating the Venetian lines with a fire-brand and set fire to the powder reserves, causing a terrifying explosion. Nevertheless, the besieging army continued its cannonade for eleven days, pounding the walls of Verona with more than 20,000 iron cannon-balls. The defenders responded with fierce fire from their cannon and culverins, well positioned in the towers and bastions of the town. The French then began digging tunnels under the walls to lay mines. But the defenders cunningly turned them to their own account, stuffing all the breaches and passages opened in the encircling walls with mines and fire-balls. When the French and Venetians attempted a fresh attack, these traps

caused them heavy casualties. Finally, the besieging army decided to concentrate all its fire on the Mantua Gate. But Fruntsperg had hastily built a second line of defence and received the French attack with massive volleys from his heavy artillery, hitherto held in reserve. The cannon were loaded with case-shot and the fragments of rough, burning metal plunged the assailants into a veritable blood bath. Finally, the French and the Venetians were obliged to raise the siege.

Artillery had again vindicated its effectiveness in static warfare, causing breaches in the thickest walls whereby the infantry could infiltrate the town. However, the defenders had been able to utilize their total resources to repulse the adversary. Castles and fortresses, hitherto deemed impregnable, saw their days numbered. Only large fortified castles, equipped with extensive artillery, could henceforth play a strategic role and, occasionally, force the enemy to alter his plan of campaign. Artillery thus dealt the death blow to the military strength of the nobility, who had hitherto resisted most sieges in their fortified castles.

This evolution resulted in a logical adaptation of architectural concepts. High, defensive walls, bastions and redoubts were thenceforth designed to provide maximum protection for the defence batteries.

The Italian wars had actually determined the entire land strategy of the sixteenth century. Infantry, cavalry and artillery combined forces into united, effective tactics. Artillery was no longer merely a psychological weapon but was becoming increasingly effective and deadly. Its fire broke up the closed lines of infantry, forced the military commanders to change their strategies, imposed new rules and tactics. Armies were learning how to deploy reserve units, to attack on the flank and to carry out elaborate manoeuvring in which artillery played its part.

The day of frontal attack, its success depending entirely on the strength, daring and courage of the foot-soldiers, was over for a time. All countries and armies adopted artillery units, with the exception of Italy which had an aversion to fire-arms.

FROM MARIGNANO
TO THE THIRTY YEARS' WAR
1515 - 1648

The battle of Marignano in 1515 and the death of Emperor Maximilian I marked the beginning of a new era. For the next three decades, France and the Habsburgs were to wrangle for European supremacy while England and Spain fought for the freedom of the seas. Since 1519 the House of Habsburg had enlarged its empire by inheriting Spain, Austria and the United Provinces, which now form the Netherlands. Since the sack of Rome in 1527, Italy had been finally eliminated from the European power struggle, while England was dissociating herself from the Continent and, more and more, turning her attention to the conquest of the seas. In France and Germany, the Reformation gave rise to serious domestic troubles without, however, diverting the Emperor and the French King from their principal objectives. Only the Turkish thrusts into the Balkans reminded Europeans of the existence of a third great power to be reckoned with.

Under Maximilian's influence and as a result of the battle of Marignano, artillery as a unit had been able to show its mettle. Drawing over-hasty conclusions from various successes, the crowned heads increased their artillery-parks to such an extent that they became more of a handicap than an advantage on the fields of battle. It was no longer possible to imagine war without artillery or, for that matter, an army without a considerable number of cannon. It was to be some time, however, before most armies learnt to place their guns in batteries, concentrating the scattered fire from individual weapons.

At the same time, the number of small arms was also increased and the harquebusiers, culverin bearers and musketeers, being much more mobile than

cannon, became a decisive element in warfare. For some time to come, infantry remained faithful to its tactics of dense, bristling formations, copied from the Swiss but, under the constantly-increasing fire-power from small arms and cannon, foot-soldiers were compelled to adopt a more flexible battle order and were more dispersed in the field towards the end of the sixteenth century. Cavalry, being particularly vulnerable to gunfire, now confined its action to the protection of the flanks.

After 1520, and for several years to come, the development of artillery was primarily motivated by its proven effectiveness in static warfare and in destroying fortifications. German towns and local princes used artillery to neutralize the belligerent and pillaging gentry. Artillery was used with notable success against Duke Ulrich of Württemberg as early as 1519. The cannon of this period were not particularly well suited to concentrated deployment. During the siege of Hellenstein, for instance, the explosion of a cannon decapitated three men; at the siege of Hohenasperg, the explosion of a double cannon, one basilisk and another piece killed two men. Undoubtedly, these accidents could be attributed to the decrepitude of the guns, whose age varied as much as their size, and to the irregularity of their calibres. Nevertheless, in spite of these setbacks, the fortress of Hohenasperg, hitherto deemed impregnable, was taken after a siege lasting ten days.

Under Maximilian's cannon, the same fate befell Hohenkrähen Castle in 1517 while, in 1523, artillery broke down the Castle of Landstuhl, fief of the famous knight, Franz von Sickingen. In one day alone, 600 shots were fired against the walls of Landstuhl which, although over 13 ft thick, could not stand up to the cannonade.

In this same year, 1523, the armies raised by the towns and the princes demolished 23 castles belonging to the knights of Franconia who were caught under a considerable weight of fire from 4 basilisks, 3 cannon, 4 culverins, 10 demi-culverins and 11 serpentines. Matthieu Harder, an Alsatian by birth, was the Master of Artillery and responsible for this impressive park. He commanded 33 Master Gunners. In 1554, the fortress of Hohenlandsberg, under siege from the troops of Nuremberg, held out 4 days against 708 cannon balls, fired by 26 cannon which the aggressors had set up over 300 yds away from the walls. Despite a complement of 56 cannon which made up its defence, the fortress was finally compelled to surrender.

During the "peasants' revolt" which shook Germany in 1525, field artillery asserted itself victoriously on all the battle-fields, despite the awesomely superior numbers of the armed peasants. In Franconia, 30,000 peasants were defeated by 5,000 lansquenets, 1,000 harquebusiers from Flanders and a modest array of artillery composed of 2 siege cannon, 2 culverins, 8 demi-culverins and 12 falconets.

From 1520 onwards, armies never took to the field without artillery support in proportion to their strength. A *Book of War*, dating from 1522, recommends the following artillery for an army of 10,000 foot-soldiers and 1,500 cavalry: 4 basilisks, 12 cannon, 2 stone-throwing cannon, 3 mortars, 1 fire-box, 16 culverins and 16 falcons. In his manual on military organization, Michel Ott of Achterdingen, Master of the Imperial Artillery, advocated a force of 55 wheeled cannon for an army of between 20,000 and 30,000 men.

The creation of large parks of field artillery was the direct result of lessons learned from action in the field on Italian soil between the French soldiers and the Emperor's troops. At the battle of La Bicoque in 1522, the Swiss mercenaries, carried away by their hatred of the Imperial lansquenets, did not wait until the French artillery was ready to fire before they flung themselves into the attack. The Emperor's artillery, backed up by harquebusiers, so riddled the Swiss with a withering fire of projectiles that more than 3,000 of their number were killed. This was the first defeat inflicted on the Confederated Swiss by the lansquenets or, to be more accurate, by their cannon, since most casualties were due to cannon balls rather than the harquebusiers' fire.

It was at Pavia, on 24 February 1525, that the Emperor won the decisive victory. A first attack was launched by 5,000 Spanish infantry who, slipping under the cannon balls' trajectory, succeeded in infiltrating into what happened to be a large fun fair that included a zoological garden. The French artillery did, however, manage to suppress any attempt at organized attack until the Spanish captured the French cannon, thus opening a fatal breach through the centre of their defence system for the German lansquenets. This battle sounded the knell of the armoured cavalry, the pride of the French nobility. The entire French artillery, including thirty-two heavy cannon, fell into the hands of the enemy. The Spanish, armed with long harquebuses which could bring down horse and rider with one round, were a deciding factor in this battle; the fact of being able to dodge the French artillery, by slipping under the trajectory of the cannon fire, gave rise to keen controversy. A further increase in artillery strength was recommended on the one hand while, on the other, suggestions were made to decrease it while increasing the rate of fire, flattening out the trajectory and concentrating fire so as to prevent the enemy taking advantage of dead angles.

The Emperor Charles V chose the second alternative, being already convinced of its wisdom by his experience at Pavia and lessons drawn from the Siege of Vienna in 1529 by the Turks, the finest artillerymen of that time. The Turks had engaged the most able Master Founders from the West. In 1522, in less than six months, they had captured the Citadel of Rhodes, considered to be impregnable, from the Knights of St John. The Turkish victory over the Hungarian cavalry in 1526 at Mohács had opened the gates of Hungary to their army while, as early as 1514, their powerful artillery had enabled them to annihilate the Persians in the Battle of Tabriz. The Turkish artillery was at the height of its power in 1529 during the Siege of Vienna, personally conducted by the Sultan Suliman II. At that time, it consisted of 300 cannon, accompanied by 120,000 foot-soldiers and cavalry. The fall of Vienna would have given them access to central Europe. To meet this threat, the defenders of this Austrian city were armed with 72 heavy cannon and 60 howitzers. Ten bastions were each mounted with 2 culverins and 6 falcons, and ramparts, each provided with 18 light cannon, extended between the raised bastions. Assailants and defenders alike made considerable use of mines and countermines during the close action. Despite intensive fire from his 300 cannon and after 20 fruitless attacks, the Sultan gave the signal to retreat. This extraordinary display of force had proved powerless against a town so well-equipped with artillery. The failure of the Turks' heavy artillery at Vienna was one of the reasons that prompted Charles V to concentrate on small and medium cannon for his armies.

Great progress was also being made in the construction of fortifications, particularly in Italy. Against these ever heavier and stouter defensive works, dispersed fire was not particularly effective. To break through such defences, gunfire had to be

Apart from cannon, during this period gunners also used mortars, capable of firing a stone or metal projectile on a high arched trajectory, which could be dropped behind the walls of enemy fortifications or entrenchments.

In the mid-sixteenth century, artillery was widely diversified. The heaviest guns in cast iron, as above, fired cannon balls weighing 124 lbs while the smallest fired shots of only 1 lb.

concentrated on the most vulnerable points or on those of the greatest tactical importance, in order to make breaches in the walls. It was also imperative to neutralize the secondary support concentrations.

Strategists and authors of military manuals formulated new rules to adapt tactics to the evolution of armies and armaments. Artillery was an important consideration. In France, the *Instruction on Warfare* was published by Du Bellay-Langy in 1535 while, in Germany, the Margrave Albrecht von Brandenburg defined the elements of the ideal army in his *Book on Warfare* which appeared in 1555. For columns on the march, he recommended a first echelon, the advance guard or scouts, made up of 59 troopers, followed by 2,000 lansquenets, 200 harquebusiers and 8 falconets. At the head of the next column was a large escort composed of 6,000 lansquenets, 1,200 troopers and the entire field artillery. The main body of the army was to consist of 4,000 troopers, 10,000 lansquenets and the heavy siege artillery. A guard of 400 troopers was to bring up the rear.

When the army was based in an entrenched or fortified camp, artillery should be provided with extra protection by erecting barriers with its chariots and train. Brandenburg also advised artillerymen to fire each gun once every morning and evening so that this loud salute at dawn and dusk would "frighten the enemy and give good heart to friends".

Before and during any battle, according to the author, the first consideration should be given to artillery and how it could be used in the most effective possible way. He believed that the military commander who knew how to make use of his heavy artillery at the right time and at the right place, with real accuracy, had already won more than half the battle. Artillery was a very expensive item, often used unwisely. In order to prevent cannon from falling into enemy hands, the barricades of chariots were to be held at all costs by bringing the cavalry into action, thus providing cover to ensure a safe withdrawal of the artillery. Albrecht von Brandenburg was also of the opinion that artillery should remain as long as possible between the three echelons

however, was that written by the Marshal of the Imperial Army, Reinhart von Solms. Published in 1556, this work comprised eight volumes. For an army of 25,000 men, the author prescribed artillery composed of eighteen siege cannons and fifty-four field guns. He recommended that the guns should be divided into batteries and not grouped in a single formation. Light cannon, especially the 2-pounder falcons, should be in position in front of the infantry. Until the end of the sixteenth century, this work by Reinhart von Solms was to remain "the most read and the most prized of works of this kind". Its author, too, was of the opinion that artillery was the decisive element of the battle, when judiciously employed.

The more widely artillery was used, the greater became the necessity for standardizing types of guns and ammunition. As early as 1526, the Master of

In Italy, the mathematician Niccolo Fontana (1499-1557), known as "Tartaglia", was the first academician to tackle the question of trajectory but he was unable to define the theory of this curve clearly enough to apply his findings to gun-laying in practice. Artillerymen laid their guns with the quadrant, shown on the left, or in an empirical fashion by means of graduated rods, as on the right.

of infantry during a battle, judiciously spread out between the lines of footsoldiers so as to be kept out of sight. He considered that this precaution prevented the enemy from prejudging the trajectories of the guns and thus making use of the dead angles to advance. In addition, he stipulated that light cannon should be quickly brought forward in advance of the cavalry, fire one or two rounds and then retire with the same rapidity. This manual would appear to contain all the rules and principles on artillery and its deployment that were known or had been adopted during this first half of the sixteenth century. The best-known military work from this period,

Artillery Jakob Preuss had proposed one set of standards for siege artillery:

Large basilisk	calibre 9 ins
Basilisk	calibre 8 ins
Cannon	calibre 5.5 to 7 ins

The calibres for field artillery were as follows:

Large culverin	calibre 5 ins
Medium culverin	calibre 4 ins
Falcon	calibre 2.7 to 3 ins
Falconet	calibre 2 ins
Heavy mortar	calibre 14 ins
Medium mortar	calibre 11 ins
Light mortar	calibre 6 ins

An engraving by Jost Amman shows the method used to break down the walls of fortified castles. Against round towers, the gunner would try to hit the same place from different angles of fire. To attack square towers, he aimed to loosen the corner stones by cross-fire.

It seems that Emperor Charles V and his military commanders adopted these proposals. They abandoned the over-heavy basilisks and equipped their artillery with the following pieces, which appeared in a 1540 inventory.

	Calibre	Length	Weight of shot
Cannon	7 ins	11 ft 6 ins	40 lbs
Medium cannon	6 ins	11 ft 1 in	24 lbs
Large culverin .	5 ins	12 ft 9 ins	12 lbs
Medium culverin	3.5 ins	11 ft 6 ins	6 lbs
Small culverin .	5 ins	10 ft 5 ins	12 lbs
Falcon	4 ins	9 ft 6 ins	6½ lbs
Falconet	3 ins	9 ft 2 ins	3 lbs
Mortar	14 ins	5 ft	100 lbs

One noticeable feature of this list is that the Imperial artillery in 1540 was composed exclusively of moveable guns which could be used equally well for static or open warfare. It no longer constituted an excessively cumbersome burden for the army as a whole and the old type of heavy siege gun, which was mounted on a flat bed, with no provision for mobility, gradually became uncommon on the battlefields of Europe, banished by the superior tactical utility of the mobile guns. Apart from the Imperial arsenals, however, other types of cannon still remained in favour. In his *Military Manual* of 1568-1570, Veitwolf von Senfftenberg mentions the following:

Tintoretto depicts the scene outside Parma at the beginning of December 1621. Their mission accomplished, the Imperial cannon have fallen silent. A stream of troops on the attack surges through a large breach in the walls opened by the cannon. Other soldiers scale the wall with ladders while the French troops flee the town.

	Calibre	*Weight of shot*
Large basilisk	14 ins	94 lbs
Basilisk	8 ins	66 lbs
Singer	7 ins	50 lbs
Nightingale	6.6 ins	46 lbs
Cannon	6 ins	32 lbs
Large culverin	5.5 ins	20 lbs
Small culverin	4 ins	12 lbs
Falcon	3.9 ins	9 lbs
Quarter culverin	3 ins	5 lbs
Falconet	2.7 ins	3 lbs
Serpentinelle	2 ins	1 lb

The works on fire-arms published during the Middle Ages were succeeded by excellent artillery manuals such as *La Pirotechnica* by Vanuccio Biringuccio of Sienna and *The New Science* by Nicolo Tartaglia. These dealt with ballistics and provided range tables, trajectory calculations and information on the range of projectiles. Biringuccio supplied accurate details on boring methods and how to achieve regular calibres. He recommended using steel bits fixed on to chucks, which were rotated by a system powered by paddle-wheels driven by water.

The battle of Pavia, as depicted by Tintoretto in 1525. François I laid siege to the city. On the night of 23rd February, the Imperial troops tried to relieve the fortress but the French guns crushed the attack. Rashly, François I silenced his artillery by ordering his cavalry to charge. He lost the battle and was, himself, taken prisoner.

Throughout the sixteenth century, artillery was constantly undergoing technical improvements. For casting iron balls, the only projectiles still used at this period, master founders used moulds which ensured accuracy.

Large foundries were set up at Heidenheim in Württemberg, Aschau in Bavaria, and Werften near Salzburg. The gun barrels were smooth, thicker at the base and tapering slightly towards the muzzle. Ornamentation remained restrained with coats-of-arms, short inscriptions, lifting handles in the shape of dolphins and moulded breech plugs. Barrels were extended to improve the accuracy of fire. Gun-carriages with limbers appeared, much lighter than the bulky gun-carriages of the past. Two iron uprights with braces supported the barrel without hampering its movement about the axis of the trunnions. It was in 1540 that the Nuremberg mathematician Georg Hartmann invented his calibre scale, a metal rule showing the internal diameters of cannon and the corresponding weights of stone, iron and lead rounds. From that time onwards, it was no longer necessary to weigh the shots in order to determine what amount of gunpowder to use before loading the guns. For a one-pounder iron shot, calibre was 1.9 ins while for the same weight of lead and stone shot the calibre was 1.5 ins and 2.7 ins, respectively. These figures clearly indicate that it was preferable to use only iron balls since stone rounds of the same weight required an altogether different calibre.

Lead pellets were only used for the lightest guns, such as falconets and serpentinelles. On all sides, efforts were made to improve the pieces firing case or grape shot, which proved particularly effective against infantry. These were filled with small balls of lead and scraps of iron, the effect being similar to modern shrapnel. Foot-soldiers dreaded this widely scattered fire from howitzers or mortars which struck at random, just as hunters' lead shot.

For laying siege to towns, the most effective pieces were howitzers and especially mortars, which alone could wreak structural damage inside fortifications or put the defenders behind the city walls out of action. As a result, an ever greater number of mortars were produced.

In the earliest days of artillery, incendiary cannon balls were only fired by mortars. Around 1550, a German munitions manufacturer adopted this basic principle to perfect a hollow ball with an explosive charge, ignited by a detonator. Fired from a mortar, the projectile exploded on impact with the ground or any obstacle resistant enough to make the striker act on the powder charge. In actual fact, these bombs did not always explode but they can be considered as the prototypes of our modern shells. Senfftenberg, the greatest theoretician on artillery during the sixteenth century, had this to say about them: "They can well be called the Terror of the World and will make heavy cannon obsolete". His prophecy did not materialize as, for at least some time to come, mortars continued to make only a limited contribution to the general deployment of siege artil-lery. In the long term, however, Senfftenberg's pre-diction proved accurate.

A great step forward in the development of guns took place in Sussex, England in 1542. Ralph Hog of Buxted succeeded in casting guns in iron. This was an important advance, for it enabled makers to produce guns more rapidly. These Sussex gun-makers produced most of the cast-iron guns used by Drake and Effingham to outshoot the brass cannon of the Spaniards in the battles with the "Invincible Armada".

The universal effort to replace muzzle-loading by breech-loading inevitably met with an insurmount-able obstacle at the time. The bronze used for the breech-caps was too malleable. Breechloading was still retained for small calibre iron cannon which were loaded from the rear of the barrel in a single operation, powder and projectile forming an "inte-grated charge". Nobody considered these rapid-firing guns very seriously, looking on them more in the light of curiosities or works of art and ingenuity.

Temporarily abandoning the idea of breech-load-ing, the Masters of Artillery cudgelled their brains to find other means whereby the rate of fire could be increased. The uniformity of calibres, the exclusive use of iron shot and the practice of employing stan-dardized measuring ladles, also called "lanterns", which enabled quick measurement of the gunpowder required for any gun, contributed to an appreciable gain in time. The field guns of King Henry II of France fired between 15 and 20 rounds per hour while, at the Battle of Montfaucon in 1565, one gun alone fired 200 shots in 9 hours.

Mention must also be made of small arms which were introduced into light artillery equipment during the sixteenth century, when the great majority of foot-soldiers were still armed with pikes or halberds. The large harquebuses and muskets, supported on a fork, fired rounds with a calibre of c.1 in and a range of 270 yds. The troops of Emperor Charles V

The German artist Melchior Feselen (1538) has not depicted the true Siege of Alesia by Caesar in this painting but the siege of a sixteenth-century city. Entrenched behind moats and parapets, the besieging army's cannon pound the town, cut off all means of access and repel any attempts to relieve the town with reinforcements brought up from outside.

used these arms successfully not only in the defence of fortresses but also in the field. The small harque-buses fired lead shot of 0.4 in and 0.6 in calibre. The double harquebuses, *serpentinelles* in French, were mounted on wheeled carriages and the harque-busiers themselves hauled them along the paths and across the battle-fields. The calibre of these pieces varied between 0.8. in and 1 in. It is interesting to note that lighter fire-arms, which were beginning to bear some resemblance to the modern rifle and, from 1550 onwards, to the pistol, were not considered as being part of artillery equipment.

During the sixteenth century, the unpredictable political situation frequently led to the siege of towns. Fortifications were strengthened and the construc-tion of defensive works underwent considerable change. As a result, both defensive and offensive artillery had to keep pace with the improvements.

Until about 1580, the Italians had shown them-selves to be the master builders of fortified walled cities. The fortifications of Milan, Turin, Verona and Ferrara are universally acclaimed as the finest examples. The defending artillery fired from the top of round bastions and tiered casemates. Inside, several lines of defenders were formed up in echelon. Their task was to fill in any breaches and hold the assailants in check. The besieging army, for their part, constructed a complete fortification system com-prising ditches, embankments, ramps or ramparts of earth, counter-bastions and towers high enough to give their look-outs a good view of what was happen-ing inside the beleaguered town. In order to pro-tect their infantry from gunfire, sappers dug ditches and communication trenches by means of which the footsoldiers could make their way to the base of the town walls without too much danger. Both besieged and besieging armies placed all their strength and faith in their artillery and their offensive and defen-sive mines against which the infantry protected itself as best it could by digging in. The cannon were also shielded by emplacements of banks of earth or low walls. As for the gunners themselves, they took refuge behind gabions or stockades impro-vised from branches woven into basket-work, re-inforced with soil.

The engravings by Dürer, above, and Burgkmair, below, give a good idea how siege artillery was deployed and protected around 1520. Only some of the guns are mounted on wheeled carriages so the cannon, not being mobile, are shielded by wooden palisades.

The besieging army's artillery had the triple task of neutralizing the enemy cannon, destroying the advanced defence system and breaching the support lines to facilitate attack by the infantry.

As a result, armies endeavoured to make their artillery as mobile as possible, to ensure its maximum efficiency in open as well as static warfare. Emperor Charles V set the example by sacrificing the number of guns for increased mobility. At the sieges of Landrecies in 1544, Metz in 1552 and Saint-Quentin in 1557, his army had only sixty cannon. He deployed seventy at that of Therouenne in 1553 and only forty for the siege of Hasdin in the same year. Throughout the first half of the sixteenth century, it was the Imperial Artillery that set the pace and

Until the mid-sixteenth century, artillery often achieved rapid successes in the siege of even powerfully fortified towns. Protected by gabions, which were basket-work tubes filled with earth, and by a defence line of infantry, the fire from these guns easily broke down walls and towers, as illustrated above by Jost Amman. However, a radical change in fortification systems, which led to fortress artillery being increasingly effective, soon altered the situation considerably.

inspired technical and strategic improvements in all armies. As early as 1535, the Emperor had ordered the casting of 70-cwt double cannon in Malaga, also known as the "Twelve Apostles", which were considered to be the best siege weapons of the time. In 1535, the Emperor's siege artillery thus comprised double cannon, large basilisks or *scharfmetzen* firing 50-pounder or 70-pounder rounds and weighing between 54 and 75 cwt, cannons firing 18-pounder to 25-pounder shot and weighing between 25 and 45 cwt, and demi-cannons firing 24-pounder or 25-pounder rounds and weighing 36 and 42 cwt.

His field artillery comprised as follows:

	Shot	Weight
Large culverins	11-15 lb	2,000-5,000 lbs
Field culverins	12-20 lb	700-1,200 lbs
Demi-culverins	7 lb	2,400 lbs
Falcons	5- 6 lb	1,200-2,000 lbs
Falconets	2- 3 lb	600-1,000 lbs
Serpentinelles	1-1½ lb	100- 500 lbs

The calibre of the cannon was 8 ins; field culverins were 3.5 ins and falconets 2 ins. About 1550, the most widely used types were cannon, falcons and falconets.

The improvements and developments in the Imperial Artillery are inseparable from the name of Gregor Löffler, the celebrated Master Founder of Innsbruck, who flourished from 1490 to 1564.

The Italian and Hungarian Wars had greatly reduced the ranks of craftsmen-founders, employed as Gun Captains in the field. The Tyrol alone had lost Wenzel Löffler and Pierre de Milan in 1527, Pierre Burgundier in 1528, Alexander Endorfer and Alexander Löffler in 1541. It was obvious that the time had come to separate the two functions of gunsmith and Gun Captain. From then onwards, it rested with the gun founders to tailor their workshops to the level of industrial production, to meet the fluctuations of demand. Gregor Löffler was ideally suited to do this. He was as much businessman as foundry specialist, and was in a financial position to bear the production costs of even the largest orders from the Emperor. On his own initiative, he purchased the raw materials and transformed his foundries at Innsbruck into a large plant, considered to be an industrial giant at the time.

Over one thousand cannon are said to have been cast in the workshops of Gregor Löffler. He was responsible for designing the moulds for all the cannon, culverins, falcons and falconets of the Imperial Artillery and, from 1550 onwards, for several other armies as well. When Emperor Charles V went to war with the League of Sinalkalde in 1546, he had 149 cannon which had been reconditioned in Löffler's workshops. Their skilful deployment enabled him to capture 369 enemy cannon.

Gregor Löffler was the first gunsmith to put his old workshops on an industrial footing, the first Master Gunner to become an arms manufacturer. From the beginning of his career, he had refused to serve as Gun Captain in the field. After living in Augsburg for fifteen years, he accepted the very favourable terms offered by the Emperor who asked him to return to Innsbruck. He was subsequently raised to the peerage. The name of Gregor Löffler is inseparable from the industrial expansion which completely altered the character of artisanal enterprises to meet increasing demands from the mid-sixteenth century onwards. Thus, artillery can be seen to have constituted one of the decisive elements in industrial and economic evolution.

The Imperial Wars against France had been responsible for the great headway made by artillery, but mention should also be made of the battles waged by King Ferdinand I of Habsburg, the successor and brother of Charles V, against invasion by the Turks. His army of 90,000 men raised in 1532 had 32 siege cannon called "wall-breakers", 36 field guns and 8 mortars. Some 12 years later, in 1544, a new expedition against the Turks was equipped with 60 siege cannon and 80 field guns. The artillery train, composed of 9,000 horses and the same number of men, carried 15 days' ammunition supply, in other words 200,000 iron shot and 12,000 cwt of gunpowder.

The German princes were unable to match the progress made by the Imperial Army and, particularly, by its artillery. The artillery of Bavarian Duke William IV comprised sixteen large basilisks, seven

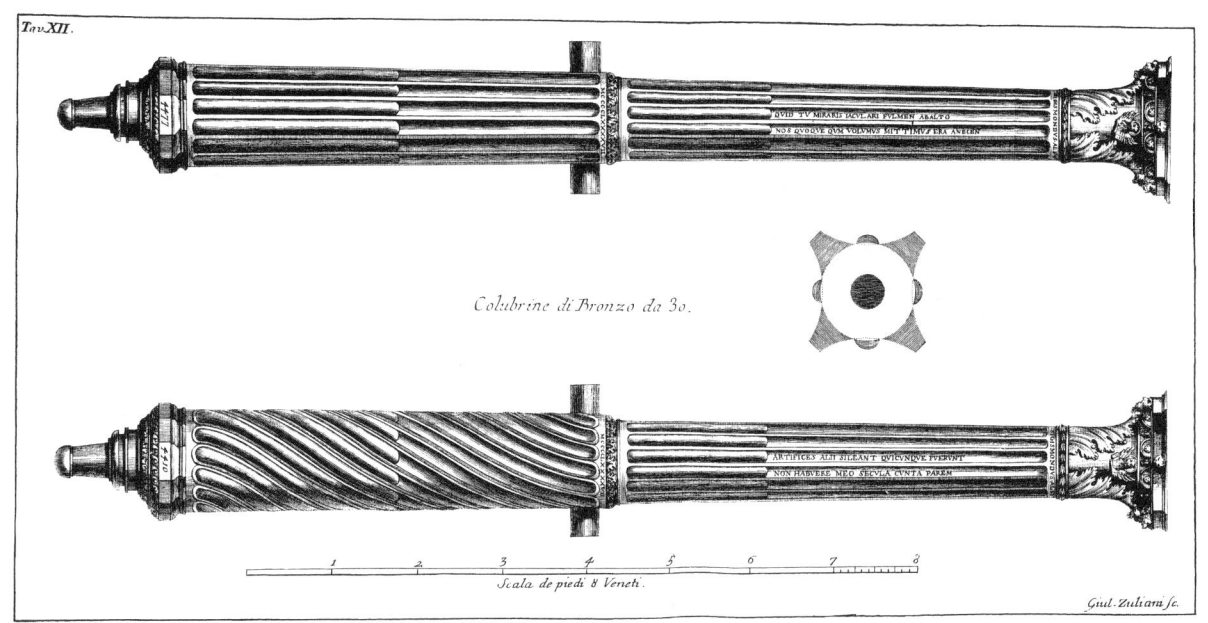

Tav. XII.

Colubrine di Bronzo da 3o.

Scala de piedi 8 Veneti.

Giul. Zuliani sc.

Tav. VIII

Cannoni di Bronzo da 12.

di Fondita Forestiera.

Scala de piedi 4 Veneti.

Giul. Zuliani f.

At the end of the eighteenth century, Domenico Gasperoni made engravings of the most outstanding pieces of artillery preserved in Venice. Some examples are shown here. Right are five 12-pounder bronze cannon and, above, two 30-pounder culverins. Below two 20-pounder culverins and, bottom, two 40-pounder cannon of German origin.

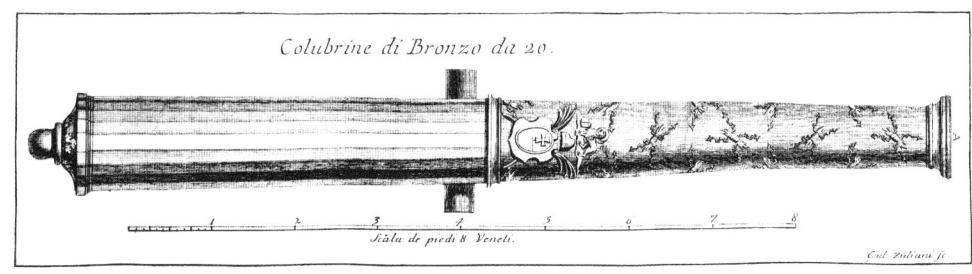

Colubrine di Bronzo da 20.

Scala de piedi 8 Veneti.

Giul. Zuliani f.

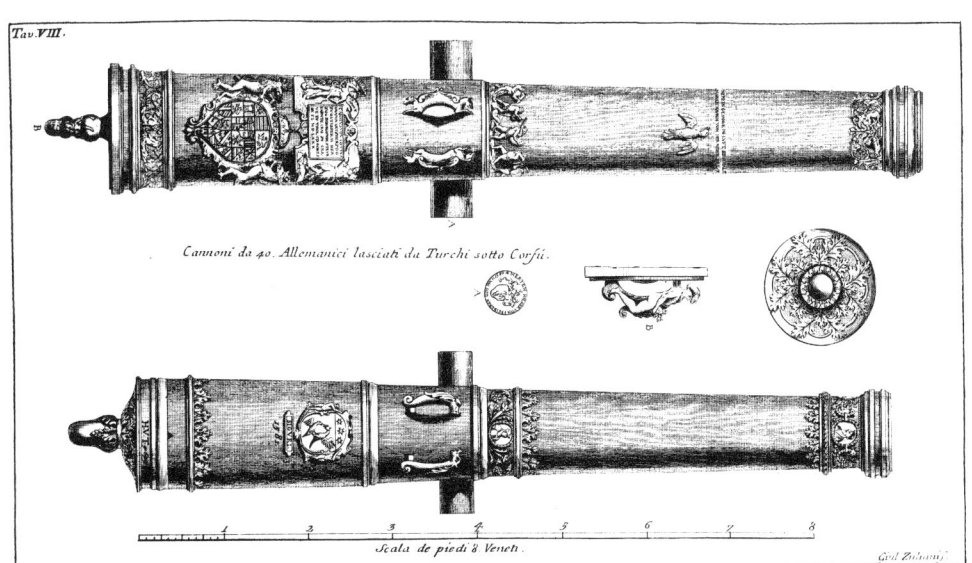

Tav. VIII.

Cannoni da 40. Allemanici lasciati da Turchi sotto Corfú.

Scala de piedi 8 Veneti.

Giul. Zuliani f.

49

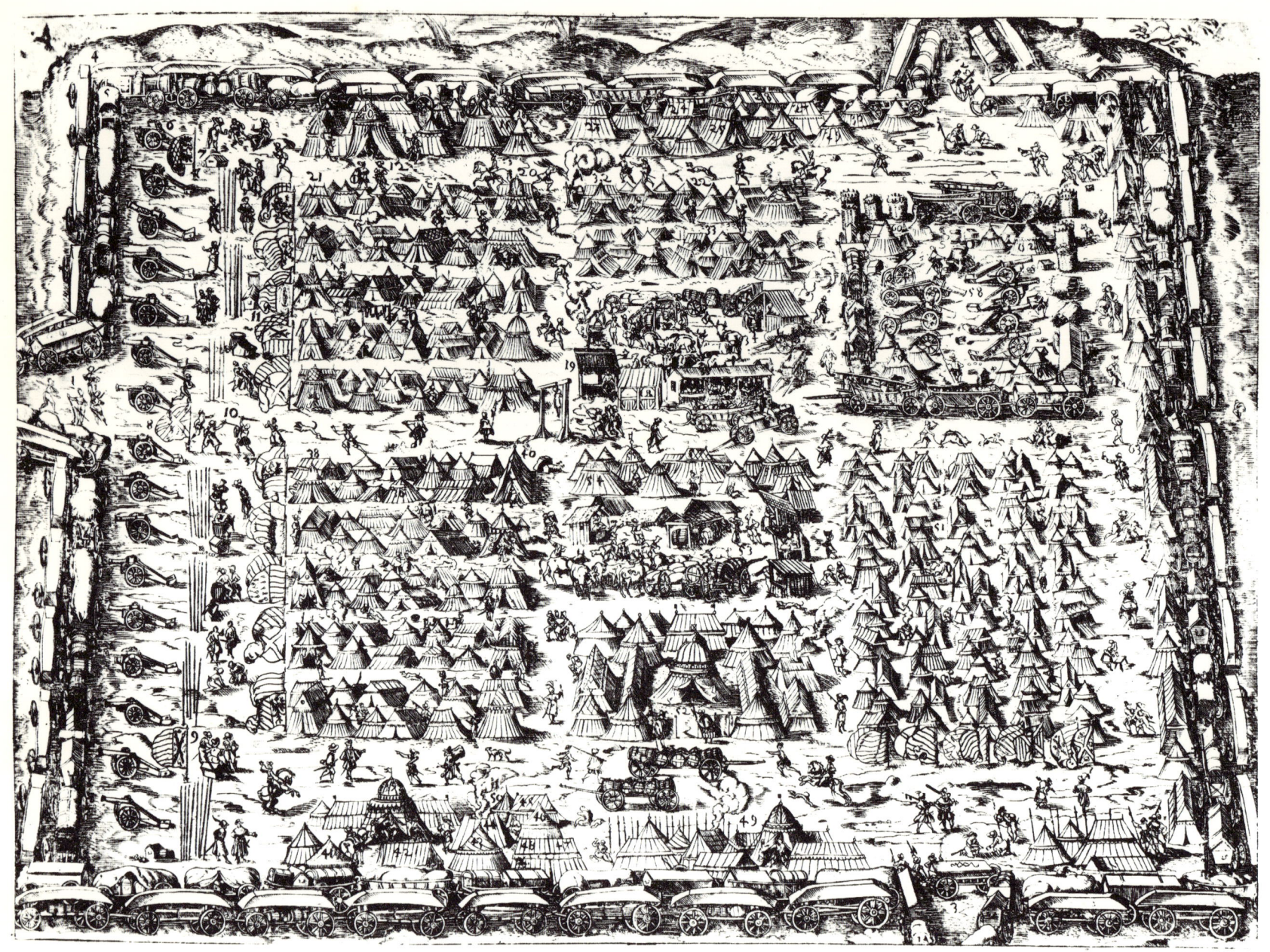

When an army set up its camp, waggons and other wheeled transport were drawn up in a huge square to form a defensive perimeter against surprise attacks, particularly from the cavalry. Some of the guns were unlimbered, facing the enemy while the rest, making up the reserve, were protected by an inner square of waggons or baggage trains, a device handed down from classical times.

nightingales, twelve singers, thirty-two culverins, thirty-two falconets and six mortars. Nuremberg, which had lost its artillery park after a succession of defeats, had it rebuilt by Gregor Löffler, the local founders being unequal to this task. Founders such as Martin Hilger in Saxony and Martin Bete in the Hesse cast a great many richly-ornamented cannon, but these pieces, with their wide variety of calibres,

proved utterly useless when matched against the modern artillery of the Emperor Charles V in 1546.

At the battle of Mühlberg, the Emperor seized 170 Hessian cannon, 131 Saxon cannon and 65 guns belonging to the armies of the towns. All these heavy pieces were richly ornamented with coats-of-arms, inscriptions in verse and anti-Papal symbols. These handsome relics from the Middle Ages were powerless against the Emperor's unostentatious but dreadfully efficient guns. Founders took the lesson to heart. After the battle of Mühlberg, they abandoned decoration in favour of greater power and accuracy, realizing that it was less important to impress the enemy by the beauty of the guns than to inflict the heaviest possible damage.

In common with the Imperial artillery, the French artillery retained almost the same length of barrel for different types of guns, the range of the projectiles depending on the calibre of the piece. The French artillery was also similar to the Emperor's in that it was primarily composed of clearly defined weaponry: cannon, culverins, falcons and falconets.

The Spanish artillery, actually forming part of that owned by the Emperor, adopted the same categories of pieces, divided into cannon, three types of culverins, falcons, falconets and sakers. An interesting feature of the Spanish guns was a double tapered barrel, thicker in the centre and tapering towards both breech and muzzle. All other European guns were shaped in the traditional way, tapering only towards the muzzle.

In a country as territorially divided as Italy, where numerous small states shared power, artillery, by reason of its diversity of types, could not compete with that of more centralized nations such as Germany and France. Nicolo Tartaglia's *Table of Calibres*, published in 1538, still mentions no fewer than twenty-six types of guns. The author, however, attributed major importance to *colubrina* or culverins with a calibre of 8 ins, cannon with a calibre of 6 to 7.8 ins, falcons with a calibre of 3 ins, falconets with calibres of 2 ins, and mortars with calibres ranging from 12 to 30 ins. The French and Imperial campaigns in Italy had brought a great number of Master Founders and Master Gunners there, many of whom entered the services of small Italian states. The Italian artillery of this period owes a great deal to foreign craftsmen and gunsmiths, although the visionary thoughts of Leonardo da Vinci and Machiavelli contributed just as much to the subsequent developments in artillery as did the Master Founders from north of the Alps. Under Dukes Alfonso II and Ercole II, Ferrara became one of the important artillery centres. In 1538, the Frenchman Pierre Greffier received the title of Master Bombardier in this city. From 1557 to 1571, the town's first founder was none other than Hannibal, son of Pierre Bourguignon, the Imperial founder (see page 27). He was succeeded, from 1577 to 1600, by Hans Lamprecht from Schaffhausen. In 1586, the artillery-park in Ferrara boasted an inventory of seventy-six guns including four culverins, four demi-culverins, four falcons, two falconets, one cannon produced by Hannibal Bourguignon, twelve whole cannon and five demi-cannon made by Hans Lamprecht.

By the sixteenth century, the era of the giant cannon had definitely come to an end. It was no longer a question of impressing the adversary but of holding him under fire from a highly mobile and quick-firing artillery. Military commanders lost all interest in large calibres, except in Moscow where, in 1586, the Tsar ordered the casting of the *Zarj Puscka* or "Imperial Crown", a gigantic gun with a calibre of 35 ins and weighing 39 tons. Admittedly, this monster's use was confined to saluting the Tsar and distinguished guests. Throughout the sixteenth century, many Master Founders from the West worked in Russia as gunsmiths or gunners, often wearing both hats. In 1521, for example, Jean Jordan from Hall in the Tyrol distinguished himself in the defence of Rjasan against the Tartars. The war against the Turks, with their extensive artillery, also compelled the Russians to equip themselves with a large artillery park.

Technical improvements in artillery could only lead to military successes in as far as the tacticians proved capable of making use of all its resources and mobility, in short, everything that made artillery the "modern" weapon of its time. In 1532, the Sultan Suliman II and his 200,000 Turks suffered a crushing defeat at the siege of the small citadel of Güns, in Hungary. The 738 defenders staunchly held out for 3 weeks against all infantry attacks and under fire from 300 Turkish guns. In 1451, the Emperor's troops laid siege to the Turkish fortress at Ofen. Driven to retreat, the Imperial infantry managed to make a safe withdrawal but the Turks captured all the artillery, totalling 36 heavy guns and 156 field guns. Never before had the Emperor's artillery met with such a disaster.

During the war of Smalkalde from 1546 to 1547 between the Emperor and the Protestant princes, the intervention of artillery had resounding effects. The two armies were face to face outside Ingolstadt on 31 August but confined themselves to exchanging heavy fire. The 6,000 rounds fired by the Smalkalde League's army did not weaken the enemy's resistance. Charles V, entrenched with his troops in a fortified camp, cried out in the thick of the cannonade: "Have you ever heard it said that a Roman emperor fell under cannon fire!" During the battle of Mühlberg on 24 April 1547, well-sustained fire from the enemy guns, far superior in numbers to their own, did not prevent the Imperial troops from crossing the river Elbe. This battle, which was nevertheless decisive, was also confined to a duel between the opposing artillery.

On the other hand, in 1550, Prince-Elector Maurice of Saxony captured the fortress of Magdeburg after a siege lasting a short time and involving a cannonade of only 1,500 shots. In 1552, the Turkish siege of Erlau in Hungary was a total failure, despite 12,000 rounds fired by the Turks' heavy cannon. The defenders aimed light guns with such accuracy that they succeeded in blowing up the heavy Turkish pieces. They rounded off their victory by making a sortie; all the guns they could lay their hands on were destroyed.

During the second half of the sixteenth century, the art of war made great strides while strategies and tactics underwent profound changes. After 1568, a new power had begun to assert itself in Europe. The military strength of the United Provinces that made up the Netherlands was a new factor which other nations learnt to take into account. In such a relatively small country, action in the field was of less significance than the siege and subsequent capture of towns. The most decisive incidents in the Dutch Wars were indeed the sieges laid to Leyden in 1574, to Antwerp in 1585, to Breda in 1590 and to Ostend in 1604. In their military books, which were the most noteworthy of this time,

Leonard Fronsperger in 1566 and Bernard de Mendoza, whose *Theory and Practice of Warfare* appeared in 1595, considered artillery as "an arm which should not be over concerned with its own safety but which should strive to achieve the greatest possible effectiveness". Both authors agreed that "even if artillery does not cause heavy damage on the battle-field, it spreads panic among the cavalry".

The Imperial artillery, mostly engaged against the Turks, maintained a very high standard. It had the good fortune to be able to rely on a very competent source of supplies, that of Hans-Christoph Löffler, the son of the famous Gregor. His foundries in Innsbruck and Vienna worked at full production:

1565 23 falcons and falconets for Salzburg

1568 12 cannon and 12 demi-cannon for the Emperor

1570 8 cannon, 4 demi-cannon, 7 culverins, 8 falcons and 20 falconets for the Emperor.

1572 4 culverins and 10 falcons for the Emperor.

1577 8 culverins, 20 double-cannon, 100 falconets, 20 falcons and 7 howitzers for the Emperor.

From 1590 to 1597, Hans-Christoph Löffler worked exclusively for the Emperor. His foundries in Vienna completely renovated the Imperial artillery park. He had perfected a light, "quickfiring" gun, breech-loaded with a cartridge. Nobody imagined for a moment that this invention would be the gun of the future and none dared to employ it in the field. It is incredible that it was merely used in firework displays to show off its marvellous mechanism. Löffler applied this principle to various guns, including three falcons in 1575 and a culverin firing rounds of between four and eight lbs, as well as two pieces of artillery each firing one-pounder shot through seven barrels in rapid succession.

Until the Thirty Years' War, static warfare predominated over open warfare, with sieges being far more commonplace than battles. As a result, several types of cannon retained the features of those built about 1550.

The new combat tactics adopted in the Netherlands by William of Nassau and Maurice of Orange were to influence the structure and deployment of artillery. From 1590 onwards, the Dutch abandoned massive formations of lansquenets in favour of regiments and companies. Regiments were composed of marksmen, and companies of pike-bearers. Troops were divided into small mobile units and their regiments equipped with complete weaponry, comprising all the types of guns in use at that period. Only heavy artillery remained autonomous. The thinner battle formations provided cavalry with the opportunity to break through the ranks of the infantry and charge with all the fury of its carabineers, pistoliers and swordsmen. Troopers and infantry no longer offered a compact target to the enemy guns. As for artillery, its missions became diversified. Some of the guns were formed in batteries for concentrated fire while the rest were deployed in support of the

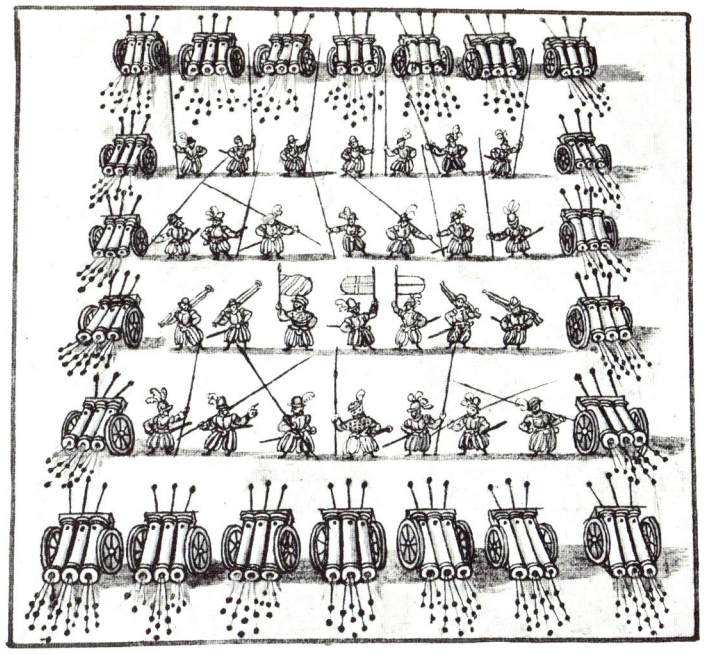

The tacticians were not unanimous in advocating that artillery should only be used in a more or less static capacity. Some were not so hide-bound. Wolf von Senfftenberg sketched out a method by which mobile guns supported infantry as they advanced. This manuscript contains some interesting illustrations on the active role of field artillery. Some of Senfftenberg's ideas were put into practice during the Thirty Years' War when Gustavus-Adolphus, the Swedish King, created his regimental artillery.

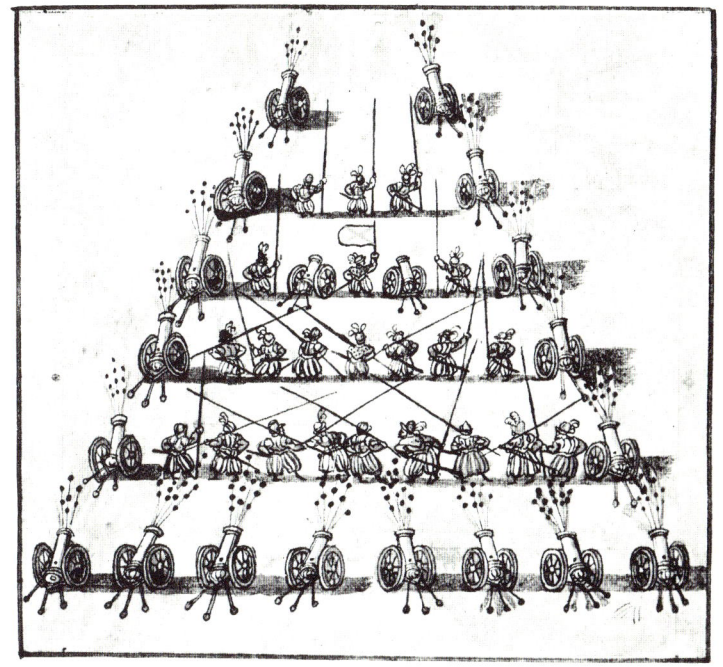

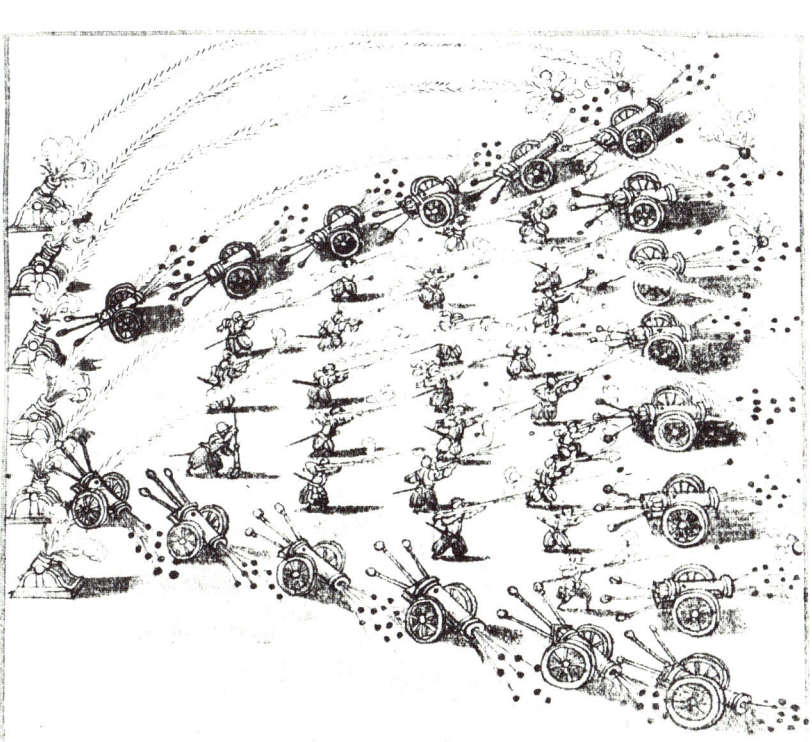

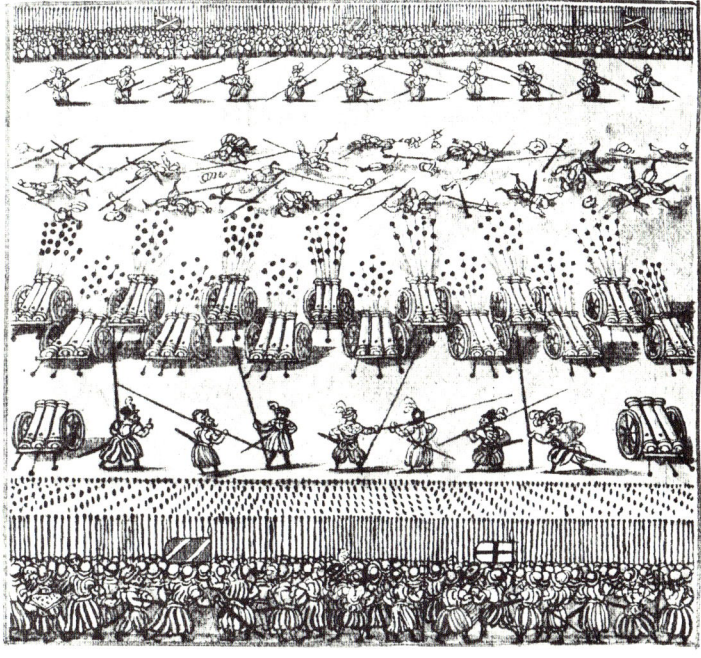

infantry. The Landgrave Maurice of Hesse advocated this form of battle array in 1601:

Centre: 10,800 foot-soldiers and 8 batteries of artillery.

Left wing: 360 troopers, 540 foot-soldiers, 4 field cannon and 4 groups of culverins firing case-shot.

Right wing: 540 troopers, 540 foot-soldiers, 4 field cannon and 4 groups of culverins firing case-shot.

The well-balanced juxtaposition of pike-bearers and marksmen gave rise to greater tactical flexibility and mobility of regiments and armies. Dutch commanders had introduced daily practice drill to maintain their troops in constant battle readiness. This training, which created the machinery for defence and attack, enabled the military commanders to have an experienced, mobile army permanently prepared. Already, an important milestone had been reached when the professional army came into being. The Dutch High Command reduced the depth of the ranks of soldiers from twenty to five and so greatly diminished the effectiveness of enemy artillery. Throughout Europe, furthermore, there was a tendency to reduce the number of guns supplied to the armies in the field. In his *Military History*, Imperial Captain Georg Fuchs recommended only 30 cannon for an army of 34,000 foot-soldiers and 6,000 mounted troops. These comprised: 9 cannon, with an ammunition supply of 5,000 rounds, 8 demi-cannon with 12,000 rounds, 6 quarter-cannon with 14,000 rounds, and 7 field cannon, or light artillery, with 16,000 rounds.

All these pieces of ordnance were expected to fire a total of 296 rounds per day, 8 per cannon and 16 per field gun.

From 1600 onwards, the following types of cannon came into use:

Cannon firing 12 to 80 lb shot.
Culverins firing 10 to 60 lb shot.
Regimental cannon firing 2 to 8 lb shot.
Howitzers firing 6 to 30 lb shot.
Mortars firing 40 to 260 lb shot.

The regimental cannon, which were actually light pieces of field artillery, replaced falcons and falconets and were assigned to the infantry for escort duty. If heavy artillery, with its cannons, culverins, howitzers and mortars, remained superior in numbers to light field artillery, now exclusively composed of regimental cannon, it was because static warfare still predominated over open warfare.

The tactics employed in laying siege to towns and fortresses had also undergone considerable change. The Dutch and the French had introduced a new system of defences combining forts and support bastions arranged in the form of a polygon. In the centre, a main tower dominated the gun-platforms. Batteries and counter-batteries were placed in a network of casemates which protected the central buildings from the outside. The Dutch displayed a marked preference for walls of earth, being easy and inexpensive to build. Although these did not provide a complete and reliable protection against gun-fire, they were relatively easy to repair or move into another position. This geometric form of fortress was to hasten the development of a parallel system of siege-works which were called into being by the necessity of approaching the fortifications under cover.

However, about 1600, on the eve of the great European wars, progress in artillery development had come to something of a standstill. On the other hand, the infantry was equipped with small arms which had been brought to a new degree of efficiency. Fresh tactical ideas had put cavalry and its increasingly well armed troopers firmly back into the saddle. As armies had become much more mobile, artillery seemed all the more cumbersome and its rate of fire ponderously slow compared to the crackling discharge of the small arms carried by the mounted or unmounted troops. The heavy costs of building and maintaining an artillery-park appeared ever more out of proportion to its effectiveness. Only heavy cannon escaped this momentary discredit, their existence justified by the ever-present need for laying siege to towns and fortified positions. For this reason alone, artillery was only to be employed as a supporting arm during the first decade of the Thirty Years' War.

54

FROM THE TREATY
OF WESTPHALIA TO THE SIEGE
OF GIBRALTAR 1648 - 1783

At the beginning of the seventeenth century, Europe enjoyed a few years' comparative calm. The Wars of Religion had exacted a heavy toll on all sides. Prompted by religious or political fanaticism, the protagonists had been ruthless and pillage was widespread. To bring down fortified points of resistance, the most effective weapons were surprise attacks, diplomacy or famine, these being employed simultaneously or successively in the course of the same siege, and were usually found to be successful in overcoming the resistance of most of the contemporary strongholds.

Early in the seventeenth century, the advance guard of an army on the march usually included a detachment of artillery. This hitherto unpublished water-colour by Captain Vasselieu, also known as "Nicolas Lyonnais", shows such a formation. Two medium culverins, firing shots of about 2 ½ lbs, flank two bastard culverins firing rounds of 7 lbs.

Alterations in tactics were scarcely perceptible, although the ideas of some of the theoreticians began to take effect. Joest Lips of Louvain endeavoured to combine the stratagems and inventions of the ancient warriors with the contemporary art of war and his theories were adopted by Maurice of Nassau, whose reforms were to help the United Provinces (the Netherlands) win their independence. At the turn of the sixteenth and seventeenth centuries, nevertheless, artillery was still considered as an auxiliary arm. No progress had been made either in its equipment, its organization or its use.

An excellent description of artillery of this period is given in a manuscript which has never been previously published. This is entitled *Treatise and drawings by which is acquired knowledge of what is observed in France concerning the conduct and employment of artillery, for the service of the very high and mighty Monseigneur Gaston de France, Duke of Anjou, only brother of the King.* This treatise was the work of Captain Vasselieu, otherwise known as "Nicolas Lyonnais", who apparently regarded artillery as a separate and self-sufficient corps of the army. The six calibres used in France were the following:

Artillery attached to the main strength of the army in battle was composed of heavy guns: four cannon, each drawn by twenty-five horses, and two large culverins, each drawn by twenty-one horses. In order of march, the cannon weighed about four tons and the large culverin three. The horses were harnessed in double file, lined up in front of the wheel-horse. In the nineteenth century, a change in the harness enabled the same weight to be pulled by only eight horses. No limbers were used, so the wheel horse had keep the trail from burying itself.

The main artillery detachment in battle formation was followed by a smaller train. The convoy comprised waggons for gunpowder, balls, followed by waggons loaded with spare wheels, the blacksmith's tool-chests, lifting gear, carpenter's and engineer's tools.

	Weight of gunpowder charge	Weight of shot	Weight of gun in shots	Weight of gun	Weight of gun complete	Point-blank range in paces	Number of horses required	Daily firing rate
	lbs	lbs		lbs	lbs			
Cannon	20	33⅓	153	5000	8420	500	25	100
Large culverin . . .	10	15¼	234	3550	6380	700	21	100
Bastard culverin . .	5	7¼	344	2500	4773	400	13	140
Medium culverin .	2½	2½	600	1500	2575	400	7	160
Falcon	1½	1½	536	800	1240	300	7	240
Falconet	1	¾	. . .	. . . .	880	200	7	200
Harquebus	1 oz.	¹⁄₁₀	. . .	45	50	120	7	300

The French cannon were equipped with trunnions and had no lifting handles or "dolphins". The cascable (breech-knob) was small. In unspecified conditions, the cannon round could penetrate twelve ft of solid ground, eighteen ft of broken soil and twenty-five to thirty ft of sand. These performances seem exaggerated, unless they were point-blank measurements. There were no mortars. The various types of gallopers were fitted with shafts for horses in tandem. Carts were used to transport the gunpowder, cannon

In a similar way to the advance guard, the rear-guard artillery travelled with a cavalry escort on its flanks. In time of war, between three and five gunners and eight to thirty pioneers were allocated to each gun, depending on its size. Gunners were unarmed and the pioneers were civilians who had been pressed into service, and who often melted away once in action.

balls, ropes and tools while the wheeled chests were reserved for the cartwrights' and carpenters' tools, for sheers, jack, axle equipment and forge. Spare wheels were fitted to an axletree and a pair of shafts to form a carriage which was drawn by four horses.

In time of war, each cannon was manned by two special gunners and thirty pioneers; each culverin had two regular gunners, two special gunners and twenty-four pioneers.

As far as the State or the Army was concerned, the administration of the artillery was completely independent, and entirely self-sufficient. A few years earlier, in his work entitled *Fortification demon-*strated *and reduced to an art*, published in Paris in 1600, Errard had given a description of guns that tallied with Vasselieu's in every respect. In 1633, the same features were described by Maximilian de Béthune, the second Duke of Sully, in a work entitled *Instruction on the facts of artillery*. This means that, for over twenty-five years, French artillery had made no progress, remaining very much the same as it had been in the second half of the sixteenth century.

The same was true of Spanish artillery. Their arsenal numbered more than fifty types of cannon. These pieces were of the culverin type. The efforts made by Charles V to reduce the number of different

Captain Vasselieu also provided details of the transport of gunpowder. Powder for the heavy guns was "granulated into small pellets" and kept in kegs of approximately 220 lbs. "Finely ground" powder was stored in kegs of 110 lbs and the priming powder in 55 lbs barrels. One cart carried from four to ten kegs or barrels, and was hurried up to the guns when they were assigned firing positions.

calibres of his guns proved ineffectual and almost as much confusion remained as before.

The lull in hostilities was brief. One day in May 1618, a handful of Protestant gentry threw the Emperor's Catholic representatives out of the windows of Prague Castle. This was the signal for a general revolt of the Protestants in Bohemia and the prelude to a war which was to last thirty years and which was to see the armies of the Elector Palatine and those of the Kings of Denmark, Sweden and France take it in turns to curtail Imperial power. At the same time, the Calvinistic United Provinces resumed hostilities against the Catholic Spaniards.

For thirty long years, all Europe remained in the throes of this conflict until the Treaty of Westphalia in 1648. Thus ended the War of Religion. The political supremacy of the Habsburgs was broken and France became for a while the dominant European power, with Sweden and Holland runners-up.

During these hostilities, technical improvements to and innovations in artillery did not originate either from France or Germany, but from Holland and Sweden. At this time, the weapons making up the artillery of the United Provinces were wholly and entirely comprised of four calibres: 48, 24, 12 and 6 cms, which fulfilled all requirements both on land and at sea. Guns of each calibre were identical, apart from occasional slight variations due to moulding faults or irregularities in casting. The cannon were provided with lifting handles, while the cascable was strong enough to be used for heavy manhandling, the most suitable position of the lifting handles and trunnions being determined by calculation. The Dutch had laid down forms which were scarcely to vary for two centuries. Another important innovation was that a weapon of each calibre was provided with a corresponding gun-carriage whose iron and wooden parts were interchangeable so that a wheel and cheek for a 24-pounder, for example, would fit all other gun-carriages of the same calibre. In addition, each type of gun-carriage had a corresponding limber which was also interchangeable. The uniformity and simplicity of this equipment was, perhaps, due to the fact that this relatively small country probably had only one centre for production. How-

ever, these innovations were not immediately adopted elsewhere, foreign artillerymen being slow to grasp all their potential advantages. In this the Dutch were ahead of their time.

However, accepted ideas and conventional practices were thrown to the winds by Gustavus-Adolphus, the King of Sweden, who entered the war in 1630. His flexible, well-constructed army had a fresh, youthful air. Instead of dense battalions, regiments were composed of 8 companies of 144 men, each consisting of 72 musketeers, 54 pike bearers and 18 false musters. The reorganized cavalry was composed of cuirassiers and dragoons. His artillery had been modernized, the 30, 16, 12, 6, 4 and 3-pounder pieces made shorter and more mobile. Depending on their weight, they were drawn by 20, 6 or 4 horses. To increase the rate of fire, the King ordered the use of light, wooden cartridges to which the shot was attached. By this means, the Swedish field guns could fire eight shots in the same time as the musketeer fired six. This artillery was divided as follows:

1) Heavy artillery for use in the siege of towns, static warfare or, in the field, when the army was crossing rivers. It often comprised two pieces for every thousand men.

2) Field artillery, following behind the heavy guns, in the proportion of three pieces per thousand men. The 12-pounders were grouped in batteries of five guns and the 6-pounders in batteries of ten.

3) Light, regimental artillery, consisting primarily of 4-pounders cast in iron. These extremely short, light guns were drawn by one or two horses or could be even carried by hand. Originally, these were the famous "leather" guns or *canons de cuir bouilli*, consisting of a tube of beaten copper, encircled with iron and covered with a leather casing, soon abandoned in favour of the iron 4-pounder which, together with its gun-carriage, weighed a total of 625 lbs. In the use of these guns in a mobile role during the course of a battle, and the moving of them as required by the dictates of action, lay the great innovation of Gustavus Adolphus. The Duke of Marlborough seems to have learned from the Swedish King, for he

certainly employed these tactics at the Battle of Blenheim in 1704. This was the first time the English used really mobile field artillery, and this use contributed enormously to their victory.

In this army, which was composed partly of mercenaries, discipline on the march and in camp was very strict. Contrary to the customs of that time, few personal belongings, no camp-followers or women were allowed in the train. Each regiment marched with its cannon and carriages and camped in battle formation in two lines. All or part of the artillery was drawn up in the centre of the first line, protected by the cavalry and squads of musketeers, and the remainder of the troops in the second line. Behind them, the train was drawn up in a square. In 1630, the Swedish King fixed a ratio of 80 guns for 12,000 footsoldiers and 85 squadrons of cavalry. At the battle of Breitenfeld, he brought into battle approximately 100 guns, 25,000 foot-soldiers and 12,000 mounted troops, the proportion of artillery being 1 piece to every 3-400 men, then a high ratio.

Facing them, the Imperial Army drew up a heavy, conventionally formed army, its basic tactical formation being the large battalion of 1,600 to 2,000 men arranged in the shape of a cross. The pike-bearers formed the centre and four squares of musketeers formed the arms. As for its artillery, Montecuculli describes it as a "chaos of material". In fact, there were few guns, c. 1/ to every 2,000 men.

Until Gustavus-Adolphus appeared on the scene, artillery had been dispersed over the whole front of armies during battle. The King of Sweden, on the other hand, concentrated his in powerful batteries in the centre and on the flanks. Some believe this to be the origin of batteries and brigades.

Until 1650, artillery-parks were made up almost entirely of cannon. At this time, the elevating screw was introduced.

A feature of this period was the reappearance of mortars, first in Holland and later in the armies of other countries. At the same time as mortars again became popular, explosive shells were finally and

Early on the morning of 12 June 1672, the French army, under the great General Condé, arrived on the right bank of the Rhine opposite a place called Tolhuis. The French artillery went into action to disperse the Dutch cavalrymen and foot-soldiers who were holding up the advance of the French Household Cavalry and the building of a pontoon bridge. Heavy action by the artillery was successful and Louis XIV, himself, was present as his army crossed the Rhine, gracing the proceedings with the Royal presence.

definitely adopted by the artillery, the use of time-and-percussion fuzes having obviated the drawbacks which had hitherto precluded their use. Later on, similar explosive shells were also fired from howitzers and cannon. The Dutch trend towards simplification and the military efficiency displayed by Gustavus-Adolphus were not, however, emulated for many years to come. Until late in the second half of the seventeenth century, the Italians remained faithful to their old types of guns while the Germans retained three different categories of fire-arms: seven calibres of cannon, seven calibres of culverins and nine calibres of mortars. The gunners and artificers did not yet form a separate military arm and were only combined in companies in cases of necessity. Usually, they were dispersed among various garrisons.

At the end of the seventeenth century, the most remarkable developments were initiated by the French. These were improvements to fortifications, the organization of artillery and equipment.

Sébastien le Prestre of Vauban (1633-1707), drawing his ideas from the Italians of the sixteenth century and, notably, from Sammichele (1484-1559), worked out a coherent and logical system of fortresses and defensive works. He advocated cross-fire by artillery, cannon to open breaches and shells to break up earth works. He introduced ricochet fire at the siege of Philipsburg in 1688 and perfected it at the siege of Ath in 1697. This indefatigable man restored 300 old fortresses and strongholds, built 33 new ones and conducted 53 sieges without one single defeat.

The first permanent corps of French artillery which, incidentally, proved short-lived, came into being at the end of 1668 in the form of six companies, four being of gunners and two of bombardiers. A few years later, in 1671, the French Corps of King's Fusiliers was established, their mission being to guard and serve the French Artillery. The somewhat misleading name of fusiliers was derived from their arms, which were fusils and not muskets. This regiment of fusiliers was combined, in 1676, with a regiment of bombardiers and, in 1679, with the first company of sappers who were attached to the artillery and under the command of artillery officers. In 1693, the French Royal Artillery Regiment replaced all the preceding formations, combining them into one. In 1710, the French Artillery had an effective strength of 697 officers and 5,630 men. In 1685, King James II of England raised a Royal Regiment of Fusiliers for the same purpose.

About this time, the French abandoned culverins and retained only six calibres of cannon: the 33, 16, 8 and 4-pounders were of French origin while two were of a type used by the Spanish, namely the 24 and 12-pounders. Unfortunately, these cannon still varied in shape and size and, worse still, in the diameter of their bores so that even projectiles of identical calibre could not automatically be used by all guns of a given calibre. Various attempts at solid casting of the guns and boring afterwards had proved unsuccessful, owing to defects in the drilling machine (cf. page 93). The range of mortars remained wider, with calibres of 6, 7, 8, 9, 10, and 11 pounds, and it seems that the trend towards simplification, apparent in the case of cannon, had not extended to these pieces. Indeed, efforts to rationalize artillery at this time encountered considerable resistance and often gave rise to violent controversy. A French ordinance of 7 October 1732 laid down new rules:

"Art. 1. Henceforth, the only cannon to be manufactured are those with calibres of 24, 16, 12, 8 and 4; mortars of 12 ins precisely and 8¼ ins in diameter; perriers of 15 ins and, for gunpowder tests, of 7 1/16 ins."

"Art. 2. The dimensions and weight of the pieces of each calibre, of the mortars and perriers, as well as the dimensions of the rings and mouldings, the position of the lifting handles and trunnions, and the ornaments of the said pieces, mortars and perriers, will remain fixed, in accordance and compliance with the tables, sketches, plans and sections drawn up by His Majesty .... without any alteration being made thereto under any pretext whatsoever."

The 24-pounder cannon weighed 5,400 lbs or 225 rounds; the 16-pounder 4,200 lbs or 262 rounds; the 12-pounder 3,200 lbs or 266 rounds; the 8-pounder 2,100 lbs or 263 rounds and the 4-pounder 1,150 lbs or 280 rounds.

This system was known by the name of the man who originated and carried it into effect, Lieutenant-General de Vallière. Not only did he reduce the number of calibres of cannon to five but he also established a fixed ratio between the thickness and weight of the pieces on the one hand and the charges and shot on the other. In 1774, in his thesis concerning the superiority of long, strong pieces of artillery, Vallière's son and successor summed up the full significance of his father's reforms. He wrote: "It was not at all arbitrarily and by conjecture that M. de Vallière decided on the important reform with which he had been entrusted. During the last twenty-eight years of Louis XIV's reign, he had observed the effects and the drawbacks of the different European artilleries and he had meditated over them at leisure during the long peace France enjoyed at the beginning of Louis XV's reign. It was on the basis of this long study that he conceived the eminently simple and productive plan of a single corps of artillery. The pieces were reduced to five calibres, from 4 to 24, which were all suited to the attack and defence of strong points and of which the first three, combined according to the circumstances, were particularly suited to field warfare so that, in case of need, strongholds could supply the armies and armies the strongholds".

In 1732, the city of Paris had nineteen guns cast with calibres of 12, 8, 4 and 2. The drawings show the structure of these guns and their gun-carriages. The barrels are richly embellished without being over-elaborate. In the same year, the Vallière ordinance stipulated certain calibres and standardized guns throughout France. Henceforth, the only guns produced were those with calibres of 24, 16, 12, 8 and 4.

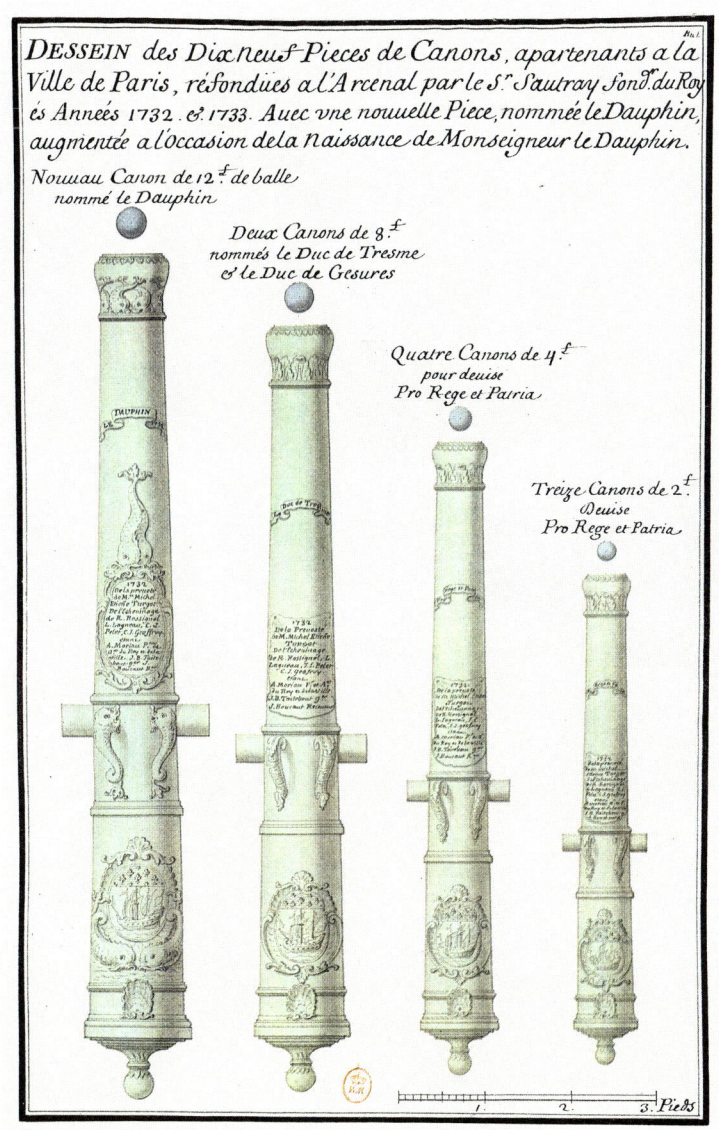

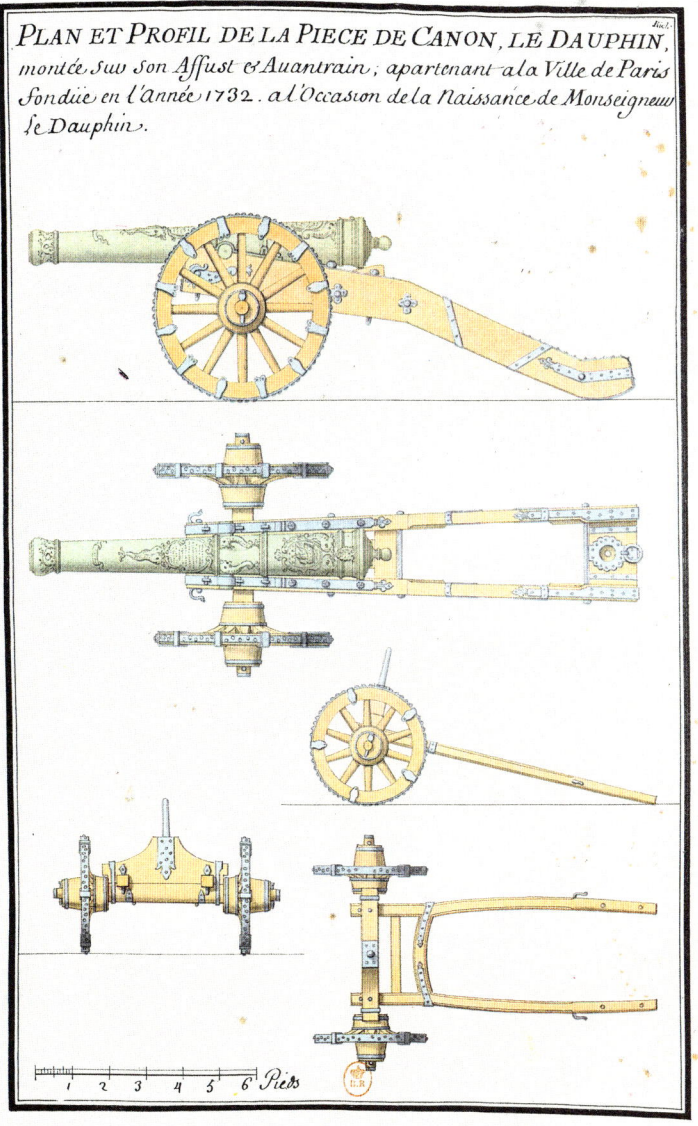

Considerable progress had thus been made towards the necessary standardization of materials. Unfortunately, de Vallière's system had not included specifications for gun-carriages and other conveyances, which still varied from one French province to another, as they had in the past. In 1745, the third edition of the *Mémoires d'Artillerie* of Saint-Rémy mentions no notable change in this respect.

At about this time, the Swedes built a very light 4-pounder weighing no more than 150 times the weight of its shot. By way of comparison, a French gun of the same calibre weighed 280 times the weight of its shot. But the outstanding features of this Swedish gun were its very easily handled gun-carriage, and an elevating screw enabling quick and accurate adjustment of the barrel to the required inclination. A small chest between the side pieces contained a ready supply of ammunition while the limber had a pole instead of a pair of shafts. Easily manoeuvred on the march, this quick-firing gun was intended to accompany the regiments of infantry and was the forerunner of field artillery. It might have been supposed that the Swedish guncarriage would arouse the admiration if not the envy of enemy artillerymen, but nothing of the kind happened. Old habits die hard and evolution was painfully slow. In spite of favourable experiments with the Swedish gun-carriage, French gunners declared it to be unacceptable and it required all the authority of the Marshal Saxe before the 4-pounders received a Swedish gun-carriage so that they could serve as escort guns for the infantry. In January 1757, a royal ordinance confirmed the *fait accompli*. When going into battle, each French infantry battalion would receive a 4-pounder of the Swedish type, drawn by three horses and followed by a sergeant and sixteen foot-soldiers.

In the mid-eighteenth century, it was the Prussians who were bringing out new ideas. On his accession to the throne in 1740, Frederick the Great found his artillery consisted of short, heavy pieces such as the carthaune, demi-carthaune, quarter-carthaune and eighth-carthaune, long pieces comprising culverins, demi-culverins and quarter-culverins and, lastly, regimental guns.

In 1747, in order to find out what kind of ordnance was currently employed by his European counterparts, Frederick the Great sent a number of artillery officers to serve as volunteers in the French army, at that time engaged in the Brabant. In the meantime, important steps had already been taken. These included the formation of field artillery, independent from siege artillery and equipped with mobile equipment, and the creation of horse-artillery which was capable of keeping up with and supporting the cavalry. Between 1742 and 1747, Frederick the Great commissioned the casting of 356 pieces, including 234 3-pounders and 60 12-pounders, thus giving marked precedence to field artillery. Later on, the King opted for heavier guns, counting on their psychological effect at a time when his enemies were reducing the calibre of their guns. His main purpose was to produce war materials ideally suited to the different tasks they were to perform and it was in this respect that the Prussian king set the lead for other powers. Moreover, he was convinced that superiority in fire power was all-decisive. In 1781, he summed up his belief in a succinct dictum: "The general's skill lies in bringing his troops close to the enemy without their being killed before beginning the attack." Following the dispositions usual at that time, his artillery was arranged in front of the centre, seldom in front of a wing and never on the flank. In this way, Frederick systematically avoided frontal assault and attacked on the flank with the main body of his troops. This was precisely what happened, for example, at the battle of Rossbach which was fought on 5 November 1757 and which lasted only an hour. The Franco-Austrian forces of the Marshal de Soubise lost 67 guns and 2,000 men, while a further 5,000 men were taken prisoner. Frederick had made a mass attack with his artillery and cavalry but had engaged only the advance guard of his infantry. He lost only 600 men.

All contemporary writers on military affairs at this time noted the importance of artillery in the Prussian army and the significant part it played in the King's successes. Frederick himself was well aware of this, as the following observation in a work which survives him reveals: "The results achieved by artillery in the last war have made it the principal element of armies." After the King's death, the Prussian artillery was to remain unchanged until 1809, when it was reorganized anew.

The Prussian artillery became the model for those of other countries and Austria, in particular, benefited. In 1762, a Frenchman, Jean-Baptiste Vaquette de Gribeauval, served with the Austrian artillery which he came to know as well as that of his own country. In a letter describing this artillery, written from Vienna to the Duke de Choiseul, he remarked that "an enlightened man could extract the best from these two artilleries (the Austrian and the French) to form another which would determine almost all engagements in field warfare". Shortly afterwards, having returned to France, Gribeauval was commissioned to reorganize the royal artillery. He based his reform on a clear-cut principle: "The weapons should be designed according to the nature of the duties they must perform." Accordingly, artillery was divided into four categories: field, siege, garrison and coastal guns. This reform affected every aspect of artillery. The length and weight of the guns were decreased while maintaining the same range and fire-power. The charge was uniformly fixed at one third the weight of the projectile. Accuracy of fire was improved by the adoption of a line of sight and the elevating screw. Rate of fire was tripled by the use of cartridge-bags, ball-cartridges and case-shot. Gun-carriages and ammunition boxes were built to a standard design.

Gribeauval encountered fierce opposition and his clashes with the champions of the *status quo* were violent. The reorganization, begun in 1765, was interrupted in 1772 by its promoter falling into disfavour, but was resumed and completed by 1774. The high command of artillery was entrusted to Gribeauval who worked unceasingly to improve his equipment until his death in 1789.

"The advantages of the new French artillery", wrote the Prussian Major W. von Graewenitz after the Revolution and the Empire wars, "had been sufficiently established during a war lasting about twenty-three years which inflamed all parts of Europe; it was consequently imitated by most other powers."

Also worthy of mention is an Austrian innovation, namely the formation of a horse artillery in 1778 by General Rouvroy. It was made up of 6-pounder cannon and 7-pounder howitzers, with some of the artillerymen being carried on the gun carriages, the remainder on the wheeled ammunition boxes. It was really more in the nature of mounted artillery than horse artillery but, with only a few variations, the formula was destined for success.

Russia was already taking an active part in the wars and her artillery was inflicting heavy casualties on the Prussian armies. The Russians had a very large number of guns and frequently used shells. Each corps was accompanied by light pieces with light gun-carriages which were well suited both to terrain and climate. The Prussian General Scharnhorst did not have a very high opinion of their howitzers, called *à la schuwalows*, and the "unicorns", chamber-cannons in the shape of a truncated cone, which were characteristic features of the Russian army during the Seven Years War. The Russian Artillery has often been conspicuous throughout the history of the Empire for its devotion to duty, and for the fact that in its ranks were found soldiers with a smattering of education.

At the beginning of the seventeenth century, the artillery of the various European military services was an incongruous medley of weapons and equipment. It was still a heavy and cumbersome auxiliary arm, used in an empirical fashion by the officers, who would sooner have relied on cavalry and infantry. Two centuries later, on the eve of the French Revolution, artillery comprised cannon and mortars with clearly defined calibres and interchangeable equipment. It had been adapted to a wide range of definite duties. Apart from regimental artillery, often handled by infantrymen, gunners and their guns formed a class apart from the rest of the army, sometimes despised by the footsoldiers and, especially, by the cavalrymen. In 1787, Scharnhorst wrote a treatise on artillery in which is found, for the first time, a logical thesis on the use of field artillery and the first views on the employment of horse artillery.

ARTILLERY EQUIPMENT
ON THE EVE
OF THE FRENCH REVOLUTION

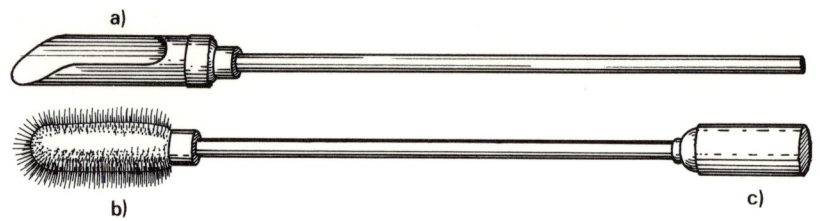

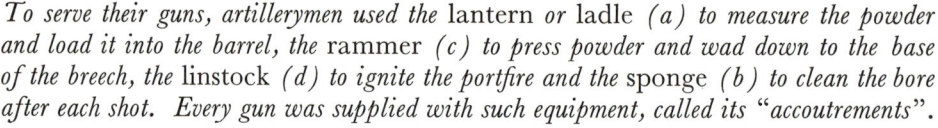

To serve their guns, artillerymen used the lantern *or* ladle *(a) to measure the powder and load it into the barrel, the* rammer *(c) to press powder and wad down to the base of the breech, the* linstock *(d) to ignite the portfire and the* sponge *(b) to clean the bore after each shot. Every gun was supplied with such equipment, called its "accoutrements".*

From the mid-eighteenth century onwards, a gun was generally defined by the weight of its projectile in lbs, a 3-pounder being a gun firing a 3-lb shot.

Each gun was made up of three parts, the *reinforces*. The back part was the *breech reinforce*, the central part the *second reinforce* and the fore-part the *chase reinforce*. The breech reinforce corresponded to approximately one third of the total length of the barrel, the second reinforce to one sixth and the chase reinforce to half. This data applied to most artilleries at the end of the eighteenth century.

The barrel was provided with *trunnions* by means of which it was supported on the gun carriage. Only the English carronades were different. The trunnions were placed near, but forward of, the barrel's centre of gravity, so as to give varying degrees of extra weight to the back or breech part. The most suitable position for these trunnions gave rise to numerous theories and experiments.

Projecting metal loops were provided through which ropes or straps were passed when placing the gun on its carriage, or removing it. These were the *lifting handles* or *dolphins* and were placed above the

gun's centre of gravity. Each gun had a cascable, which could be used for heavy manhandling.

Moulded ornamentation gave the guns a pleasing appearance and the French artillery set the fashion. The rear moulding was the *breech ring* and the front one the *muzzle collar*, being not only ornamental but also used to strengthen the muzzle.

The *bore*, or the hollow down the barrel, had an imaginary line running through the centre, the *bore axis*. The rear end of the bore, the *breech*, was of a shape which varied from one artillery to another.

There was some disagreement as to where the *priming-hole*, through which the charge was ignited, should be placed; whether it should ignite the charge at the very rear of the bore or further forward. Opinions were divided. Scharnhorst thought the ideal arrangement would be for this vent to be situated at least one inch from the rear of the bore. The diameter of the vent also varied widely from one national artillery to another; in France, it was "two lines and six points", or approximately 5 mm.

The bore diameter was greater than that of the projectile by about 2 mm. or one twelfth of an inch.

66

This clearance between the round and the metal of the barrel was the *windage*. Without this windage, the projectile could not easily be inserted into the barrel; it was all the more necessary since no foundry making cannon-balls, nor even any boring machine, could yet guarantee a constantly uniform diameter. Whether the amount of windage should vary according to the size of the gun and how its dimensions could be determined accurately were questions which preoccupied many artillerymen. Experience had shown that an over-large windage decreased both the range and the reliability of fire.

In the French artillery, Gribeauval reduced the windage by half, thereby achieving greater firing accuracy, due to the projectile bouncing about in the bore to a lesser degree. Less wear on the guns was evident, owing to the projectile having less play inside the bore. In addition, increased range was possible, as the explosive gases were not dispersed round the projectile to the same extent.

However, it appeared that artillery experts at the time did not have a very complete or profound knowledge of physical phenomena, nor of the technical processes whereby guns could be cast and manufactured with the necessary precision.

The length of a gun barrel was always expressed in multiples of either the calibre or the diameter of the projectile. Thus, a gun was eighteen calibres long when its length was eighteen times greater than the diameter of its projectile.

The question of determining the most suitable length for a gun was not resolved scientifically. The most important and systematic trials during this period took place in Hanover in 1785. It was then established that the best range was achieved by guns whose length was eighteen times their calibre and whose charge weighed half as much as the projectile. An extract from these experiments is shown below.

Average Ranges

Guns	Elevation in degrees	Length in calibres			
		24	21	18	16
12-pdr	1	953	978	982	802 Paces
	2	1348	1401	1280	1299 "
6-pdr	1/6		530	492	533 "
	1 1/6	873	925	990	829 "
	2 1/6	1285	1264	1278	1150 "
3-pdr	1/6	415	411	446	391 "
	1	748	825	810	715 "
	2	1115	1035	1150	988 "

From this table, it will be seen that guns with lengths of 24, 21 and 18 calibres, using charges weighing half that of the projectiles, achieved pretty much the same range whereas, with a length of only 16 calibres, ranges dropped by approximately 100 to 150 paces.

Prior to the French Revolution, no records are available of systematic experiments to determine the ratio of recoil with different weights of guns and gun carriages, all other factors being equal.

The weight of a gun was generally expressed in pounds, sometimes subdivided into pounds of metal per pound-weight of projectile. For example, a 12-pounder gun, sixteen or eighteen calibres in length, would have the following weights in terms of charges specified for a charge of:

3 pounds (2/8) = 100 pounds of metal per pound-weight of projectile, or 1,200 pounds.

4 to 5 pounds (3/8) = 150 pounds of metal per pound-weight of projectile, or 1,800 pounds.

6 pounds (4/8) = 220 pounds of metal per pound-weight of projectile, or 2,400 pounds.

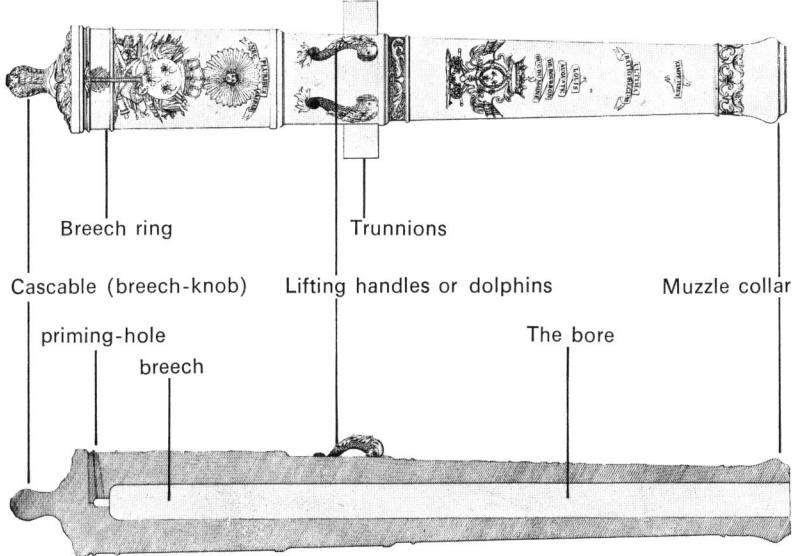

Breech ring   Trunnions

Cascable (breech-knob)   Lifting handles or dolphins   Muzzle collar

priming-hole   The bore

breech

The heaviest field guns easily weighed 1,750 lbs and it was no easy matter hoisting them on to their gun-carriages. In the field, gunners employed a system of ropes and beams for this purpose and used the gun-carriage as a lever by combining the effect of wheels and ropes. By this means, they hoisted the gun into its position, either from the front of the carriage, as shown above right, or from the back, seen left.

Guns were manhandled over short distances while, for longer distances, one gun was moved by using another or, better still, a sling-cart, as shown above on the left. The gin or sheers seen below was the hoisting appliance most commonly employed by the gunners.

These rules were based more on the results of experiment than on any scientific knowledge.

The thickness of the walls of the barrel was obviously not uniform over its entire length and Scharnhorst noted his findings in this respect:

"In most artilleries, the bore is divided into two almost equal parts, one comprising the first and second reinforces and the other comprising the chase. The decrease in thickness of metal over the first and second parts is, respectively, in the ratio of 4:7. Thus, if the thickness at the base of the bore is twenty-two twenty-fourths of the calibre it will be eighteen twenty-fourths at the middle, or rear portion of the second reinforce."

Here again, this assumption was purely empirical, the theoreticians having not yet succeeded in formulating exact equations.

Gun-Carriage and Equipment

The gun was placed on a gun-carriage, consisting of two large blocks of wood, the *cheeks*, connected by three of four other pieces, the *transoms*. The gun-carriage was mounted on spoked wheels, their rims being reinforced with iron. For some pieces of

Although artillerymen normally used horses to move their guns, they sometimes displayed considerable ingenuity by combining blocks and tackle or other makeshift winches.

A rope, the prolonge, *connecting the limber to the trail of the gun-carriage, made for greater manoeuvrability in retreat or when shifting the gun's position in action.*

Two or three men were able to place a light gun in position on its carriage. Below, on the left, the cheeks are used as a lever and the wheel, turning backwards, slides the gun into its cradle. On the right, a carriage wheel has been removed and the gunners are hoisting the gun up sideways over two beams arranged as an inclined plane, using the wheel as a purchase for their tackle.

With a frank and good-natured realism, Pierre Puget (1620-1694) depicted a gun being tested at the Port of Antibes. The gunners have pulled the gun up on to a hillock and set up the target several hundred feet away. The heavy gun has just fired; some of the men take an interest in its performance, others while away the time or quench their thirst.

ordnance such as garrison guns, the gun-carriage merely had small, solid wheels.

The gun on its carriage was provided with gun arms or equipment. These were the *lantern* or copper ladle used for putting the gunpowder into the bore; the *rammer*, a thick wooden pole used to press the plug or wad down on to the gunpowder and then on to the projectile; the *sponge* to swab out the barrel after each shot and the *linstock*, fitted with a wick, used to ignite the *portfire*; and the *portfire* and *portfire cutter*. In the field, the gun-carriage supporting the barrel was coupled to a limber consisting of two wheels, an axle and two pairs of shafts. It was not until later that pole draught, which enabled two or more horses to be used, was generally adopted. Guns were also transported on four-wheeled carts, being transferred to gun-carriages on the battle-field.

Gun Laying

About the end of the seventeenth century, French gunners knew two basic methods of gun laying. They either directed the axis of the barrel on to the target, using the highest points of the breech and chase as guides, known as *point-blank fire*, or fired at an angle of 45°. This second method of *firing at maximum elevation*, for guns, and full range gave the longest possible trajectory. There was no means of firing with any precision over distances greater than point-blank range, in other words, over an average of three hundred fathoms, or six hundred yds.

However, in other countries, artillerymen were already using the quadrant or quarter circle to angle their guns accurately.

At the end of the eighteenth century, the tangent-scale provided the gunner with other lines of sight, variable according to the distance, thus enabling gun-laying to be corrected by the observation of previous shots. This innovation was described by Scharnhorst in glowing terms: "There is not the least difficulty in adopting a sight of a shape suitable for any kind of gun whatsoever, old or new. A hole is made in the thick part of the breech, at right angles to the axis of the bore; a cylinder is placed therein, graduated on one side in power and lines of sight and, on the other, in degrees. This cylinder has a head by which it can be lifted. A screw running horizontally through the breech holds the cylinder in place when it has been raised to the required height. This construction is both so simple and so advantageous that it is to be wondered at that others could have been conceived which are not at all suitable for the purpose."

In order to make a traverse, gunners used the handspike attached to the back of the gun-carriage so as to swivel gun and carriage round to the position indicated by the gun-layer. Laying was facilitated by the elevating screw which came into use from 1650.

Charge, Projectiles and Firing

Loading the guns was a tricky operation. The measure of gunpowder and projectile were inserted into the barrel through the muzzle. But before this, it was essential to ensure the barrel was cleared of any residue from the previous firing or, for that matter, of any dust which could have found its way

In the Vallière system, all the dimensions were laid down rigidly. However, standardisation was difficult to achieve in practice, for the measuring instruments of the time lacked precision. Vallière 24–, 16–, 12–, 8– and 4-pounder cannon are shown below.

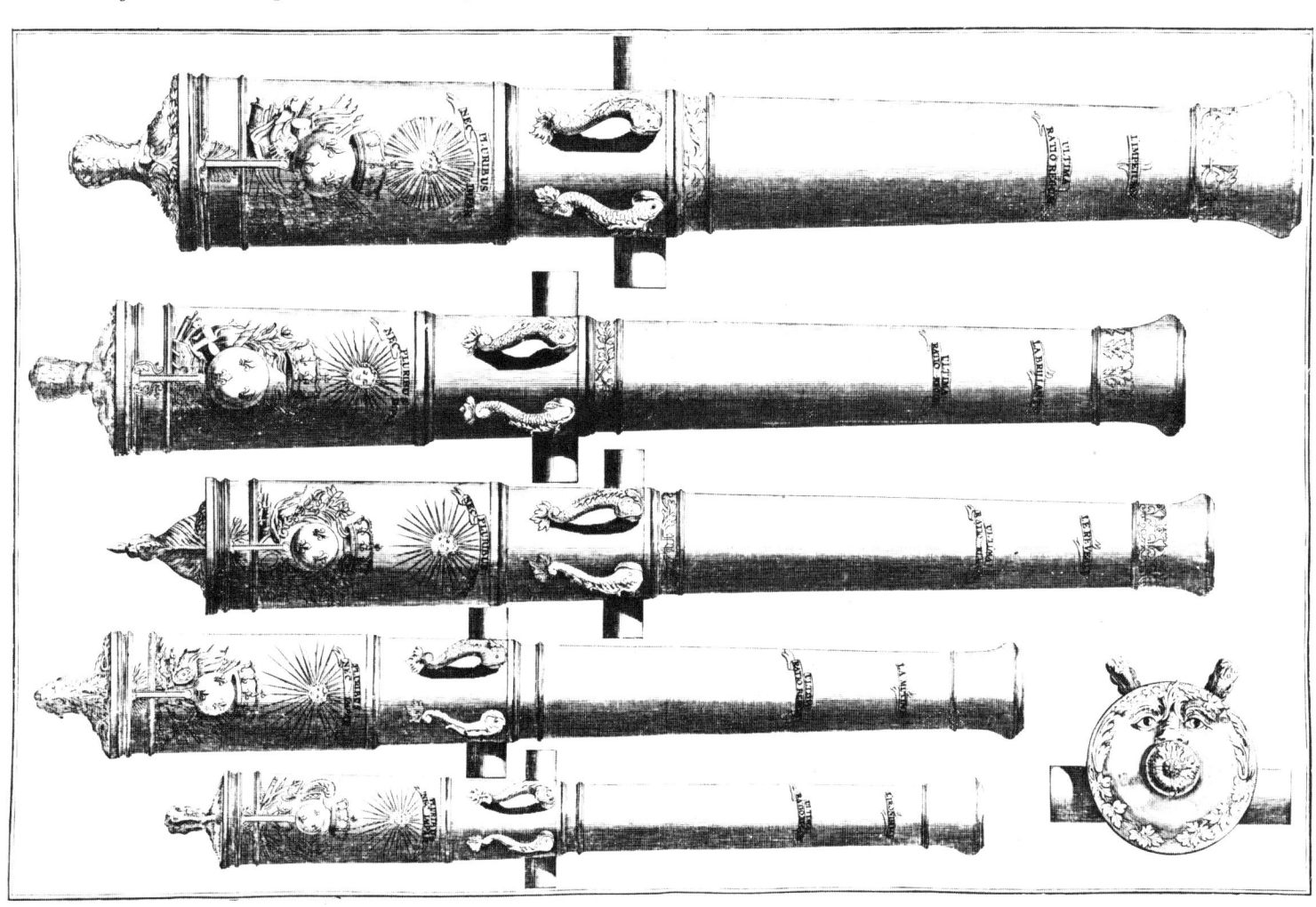

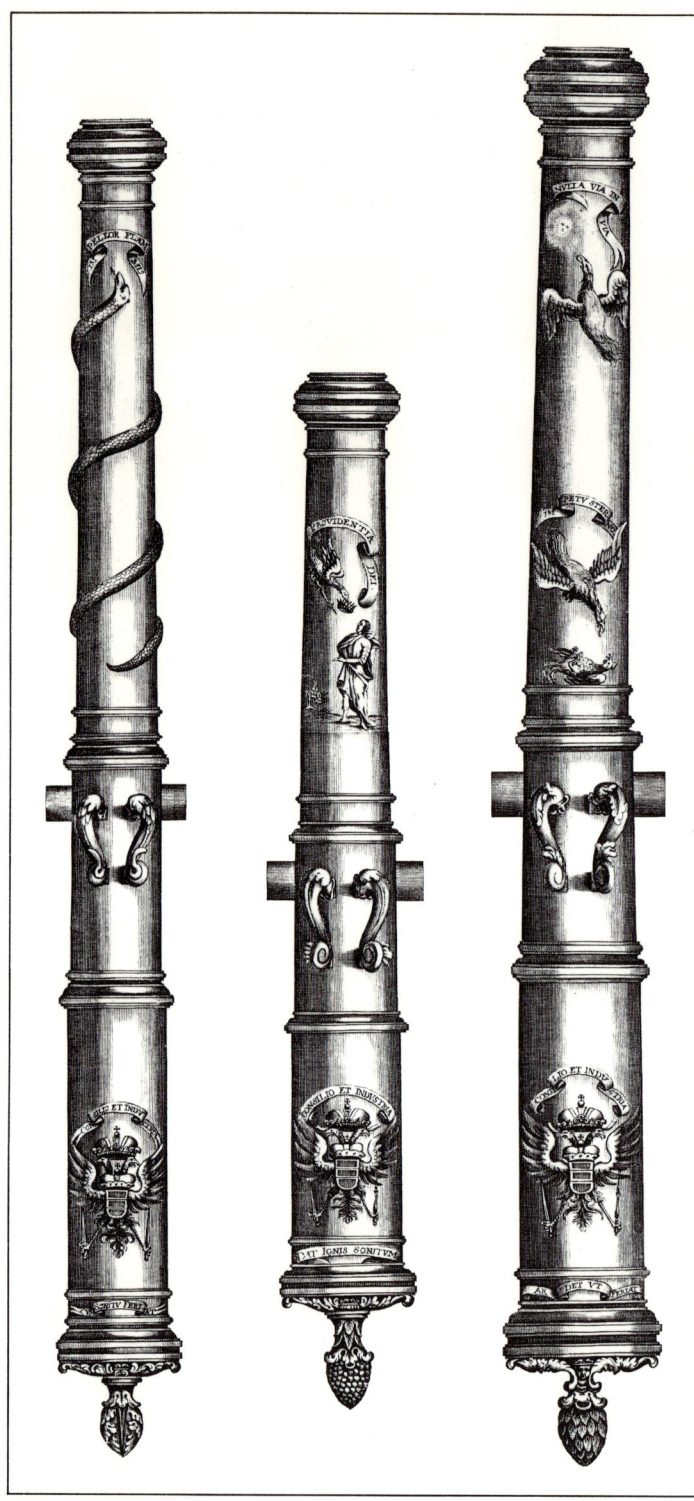

During the first half of the eighteenth century, Austria retained two types of long guns: cannon or Karthaune *and culverins or* Feldschlangen. *The cannon had a proportionately shorter barrel than culverins, that is fifteen to twenty times greater than the calibre, the barrel length of culverins being about thirty times that of their calibre. A 9-pounder demi-culverin can be seen on the left, a 12-pounder quarter-Karthaun in the centre and, right, a culverin firing rounds of eighteen pounds.*

into the bore on the march or while on the battle-field. A long-handled sponge was used for this purpose. Then, by means of a large cylindrical ladle, usually made of copper, the gunpowder was poured into the breech, at the base of the barrel. In order to compress the gunpowder into as compact a mass as possible, a wad of straw, hay or tow was then inserted and pressed down with a rammer. The projectile was then placed in the bore, followed by another wad. Finally, the whole mass, composed of gunpowder, wad, projectile and second wad had to be rammed down as tightly as possible to ensure satisfactory firing.

It will be remembered that Gustavus-Adolphus had used wooden cartridges attached to the projectiles to increase the rate of fire. A whole century passed before this method was adopted in other countries. The cartridge-bag only came into use towards the end of the eighteenth century. The powder charge was contained in canvas, parchment or cartridge-paper bags. Gribeauval stipulated that the bags should be made in twilled serge while Sharnhorst favoured wool.

The charge was described in relation to the weight of the shot; for example, a charge of $1/3$ weighed one third as much as the projectile. Therefore, the gunpowder charge would weigh four lbs for a 12-pounder gun, although this proportion obviously varied depending on the range and the angle of fire. The gunpowder was made up from seventy-five parts of saltpetre, twelve and a half parts of sulphur and twelve and a half parts of charcoal. Its composition scarcely varied from one gunpowder factory to another, or even between countries.

Igniting the charge was a relatively simple business, carried out through the priming-hole or vent. This was filled with finer gunpowder which trickled down to the gunpowder charge. Using a lighted portfire, lit from the slow-burning linstock, the gunner would ignite the gunpowder in the vent which, in turn, ignited the charge.

Cannon normally fired solid shot, well-rounded and free from rough edges so as not to scratch the bore. They were often retrieved after a battle and used again.

Hollow projectiles were sometimes used. These were tin canisters filled with lead balls, nails or other types of case-shot. There was also grape-shot: "wooden plates with a wooden core in the middle around which a great number of lead shot were stuck with tar or pitch". During a siege, gunners often fired red-hot projectiles to set defences and houses alight. This type of shot was heated on a charcoal fire until red-hot. In this case, loading was carried out with the help of tongs and wads were made of fired clay, rather than inflammable straw. It was estimated that to prepare the fire and bring it up to the required temperature, six quintals or hundred-weights of charcoal were needed, as well as a further twelve lbs an hour to keep the fire going. On a well-prepared fire, a 36-pound shot took thirty to thirty-five minutes to turn cherry-red while the actual loading of a 36-pounder subsequently took another whole minute.

Serving the Guns

In the field, artillerymen had to be able to carry out a certain number of essential operations, namely:

1. Loading the gun, laying and firing it.
2. Moving the gun on its carriage forwards or backwards and, if necessary, lifting it off the gun-carriage and replacing it.
3. Removing and replacing the limber.

Each artilleryman should normally be able to carry out the duties for which he had been trained, regardless of the type of gun. The procedure for gun loading was as follows:

Gunner A takes up his position on the right, near the gun muzzle. He swabs out the gun with the sponge and pushes the cartridge-bag to the base of the bore. When the gun is fired, he retires, making a quarter-turn to the right.

Gunner B is stationed on the left. He inserts the cartridge or the gunpowder, using the ladle.

Gunner C points the gun by means of the hand-spike on the back of the gun-carriages. He has at least two linstocks and some portfires. He stands at the back of the gun, on the right.

Gunner D plugs or "serves" the vent as Gunner A is withdrawing the sponge. He punctures the cartridge-bag with a pricker, applies the quick-match or portfire to the priming-hole and lays the gun. He stands behind the gun, to the left. The most intelligent gunner was selected for this task.

Depending on the calibre of the cannon, each of the gunners needed one or several assistants. A non-commissioned officer or senior soldier was usually detailed for the duty of Gun Captain. His responsibility was to see that all operations were carried out correctly.

Some of these tasks were quite dangerous. Gunner A could be burnt while swabbing out the gun and, sometimes, bad coordination even resulted in the accidental loss of an arm. This particular job was scarcely coveted.

Removing and re-hitching the limber were operations which took time and required considerable strength. Gunners facilitated handling by inserting a lever into the muzzle and pulling on the wheels. During battle, the limber was moved to a distance of at least twenty paces when it carried ammunition.

In action, a well-trained artillery force could be advanced, firing two shots in less than two minutes at every 150 paces. In other words, by sheer brute strength they could advance 300 yds, firing four shots in approximately four minutes. When falling back, the use of the ammunition waggon fixed to the

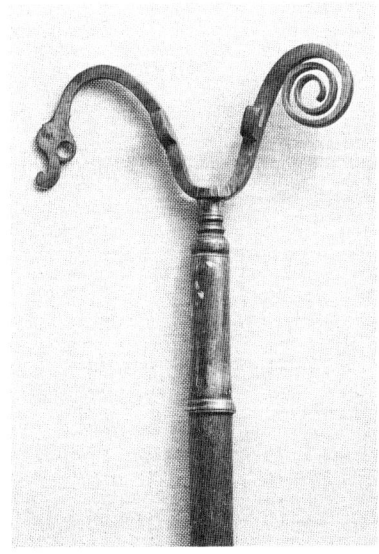

For each gun, at least one artilleryman was responsible for a linstock, a rod about six feet long, one end fitted with a holder for the slow-match which was kept alight throughout the course of a battle, the other pointed, to be stuck into the ground between each pair of guns.

The mortar was directly derived from the perrier and used to reach the enemy sheltering behind city walls or taking advantage of undulating ground. It fired explosive shells in a highly-arched trajectory. It was gradually superseded by the howitzer.

into the muzzle. Twelve men divided among these three levers and walking on each side of the cannon could transport 1,800 lbs or, theoretically, any field gun laid down in the Gribeauval system, for example. In practice, this system of levers and ropes was also used to hoist the gun up by its side on to the gun-carriage, one of the wheels having been removed.

According to Scharnhorst, sheer legs could also be employed. A rope was run through a block and tackle, the hook at one end being attached to the gun. The other end of the rope was wound round a winch, turned by means of hand spikes. By this means, a 24-pounder gun could be lifted off its gun-carriage or replaced.

The choice of method depended solely on the officers and no rules were laid down for such handling by brute strength. A great deal of time was spent to the accompaniment of loud oaths and descriptive invective, for which no provision has ever been made in any of the regulations.

Mortars and Howitzers

At the end of the eighteenth century, the perrier had been superseded by the mortar which was not designed according to any established theories or systematic experiments. Even with mortars of equal calibre and using the same powder charge, length and weight varied widely from one country to another. In France, the length of the mortar was one and a half times its calibre and, at this period, a mortar was never longer than two calibres. Vallière's ordinance of 1732 stipulated mortars of twelve ins in diameter and eight and a quarter ins in calibre. There were three types of twelve-inch mortars. One had a cylindrical chamber containing five and a half lbs of powder, another had a pear-shaped chamber which also contained a charge of five and a half lbs. The third type had a pear-shaped chamber holding twelve lbs of powder. The first two types were intended for medium-range firing, over a distance of 1,500 yds, while the third was capable of longer ranges, up to 2,300 yds. But these mortars wore out very rapidly, after only seventy shots for the first and second types and after only twelve to twenty rounds for the third.

back of the gun-carriage enabled gunners to cover a distance of 300 yds while firing four shots twice in four minutes. Obviously, such manoeuvres were carried out in great haste but it was considered that gunners should be capable of moving their guns over favourable ground at the rate of 150 paces per minute without undue strain.

Occasionally, a gun had to be moved a short distance, as when a gun-carriage had been damaged and required replacement. At such times, a rope was tied round the cascable and a lever passed through the rope at right angles to the axis of the gun. A second lever was placed through the dolphins. A third was arranged crosswise under a fourth driven

Shorter than the cannon, the howitzer was composed of a breech reinforce, a second reinforce and the chase.

This cross-section shows a howitzer loaded with a fuzed round. The chamber has been successively filled with powder, a wooden wad inserted and the bomb positioned with its fuze facing outwards. The gunner lit the fuze, then the charge.

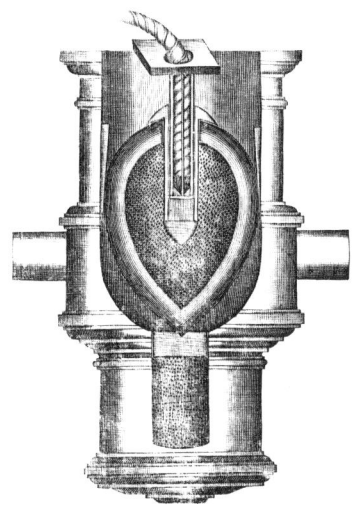

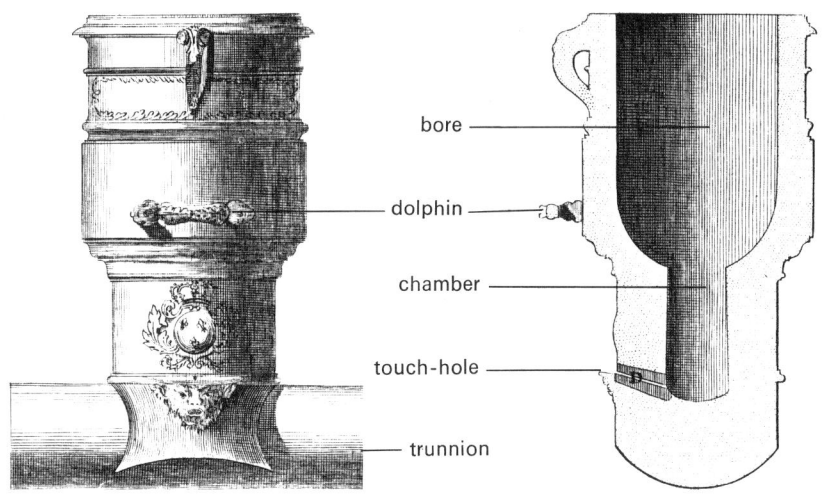

bore

dolphin

chamber

touch-hole

trunnion

The mortar was shorter and broader in shape than the howitzer. Its trunnions, placed at the blanked-off end, rested on a different type of gun-carriage, called the bed. The projectiles it fired generally weighed from 10-60 lbs.

The mortar in the picture below has been set on a temporary bed for a test shot. The acceptance of such guns was subject to various trials and inspections. If the gun was cracked or defective, it was rejected. On the right of the picture, the gunners are preparing a howitzer which will be subjected to the same gun trials.

Efforts were made to replace these with a mortar of twelve ins in calibre, capable of firing over long distances with a powerful charge. Experiments were carried out with different shapes of powder chamber, various alloys as well as varying thicknesses of metal. The results were so unsuccessful that the idea of producing twelve-inch mortars, capable of firing 150-lb projectiles over a distance of 1,200 fathoms, was completely abandoned. Instead, artillery experts chose a smaller calibre, the ten-inch mortar. Its metal was almost thicker than that of the twelve-inch mortar, and it was crammed with seven lbs of gunpowder. During trials, it was found that this mortar could fire 300 shots before having to be taken out of service while its effect was very much the same as that of the twelve-inch. The Gribeauval ordinance called for mortars of twelve, ten and eight ins. During the same period, the other countries were making similar experiments.

Mortars fired bombs, or hollow projectiles, filled with gunpowder which exploded after a given length of time.

In 1646, Malthus recommended a method of aiming mortars which was to remain in favour until the mid-nineteenth century. This involved placing a quadrant on the front face of the muzzle. A shot was fired at an angle of 45° with a first charge. If the shot over-reached the objective, he modified the gun's elevation to a steeper angle. If, at an angle of 45°, the shot fell too short, the charge was increased.

Howitzers

Another type of gun made its appearance on the battle-fields at this time, the howitzer, a piece of ordnance halfway between the cannon and the mortar. The howitzer was derived from the old-fashioned perrier which had proved too heavy and

Red-hot cannon balls were fired day and night during a siege, with the intention of setting the enemy's stores alight. As in the illustration below, the shot was heated until cherry-red. A powder charge was inserted and rammed down, followed by a wad of damp hay or clay. The gun was aimed, the rounds dropped down the barrel with tongs and the gun fired immediately to avoid the heat setting off the charge prematurely.

Der neue 8℔ige Haubitz.

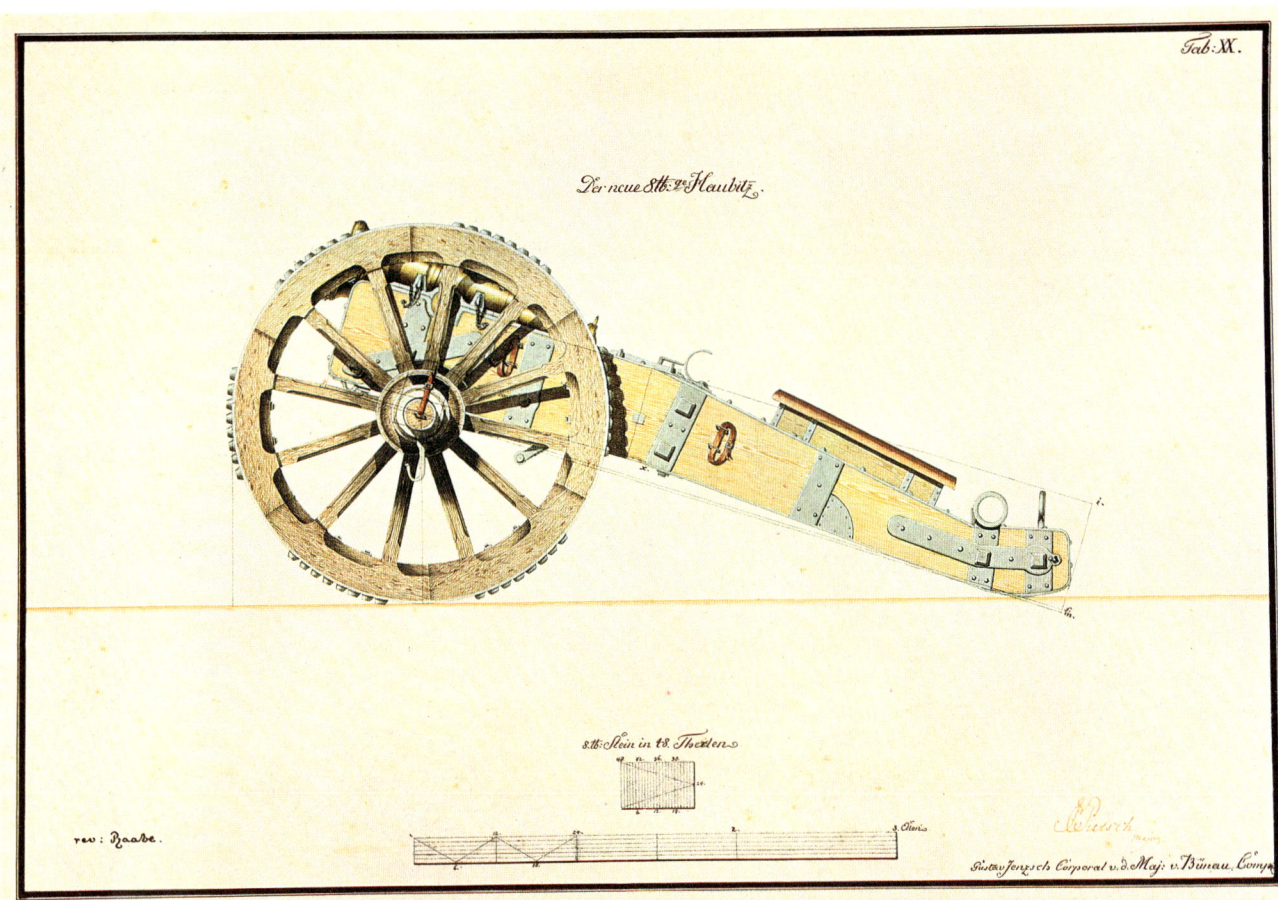

8℔ Stein in 8 Theilen.

Gustav Jentzsch Corporal v. d. Maj: v. Bünau, Comp.

Der Mortier-Wagen eines 32℔igen Block-Mortiers.
Fig: 88.

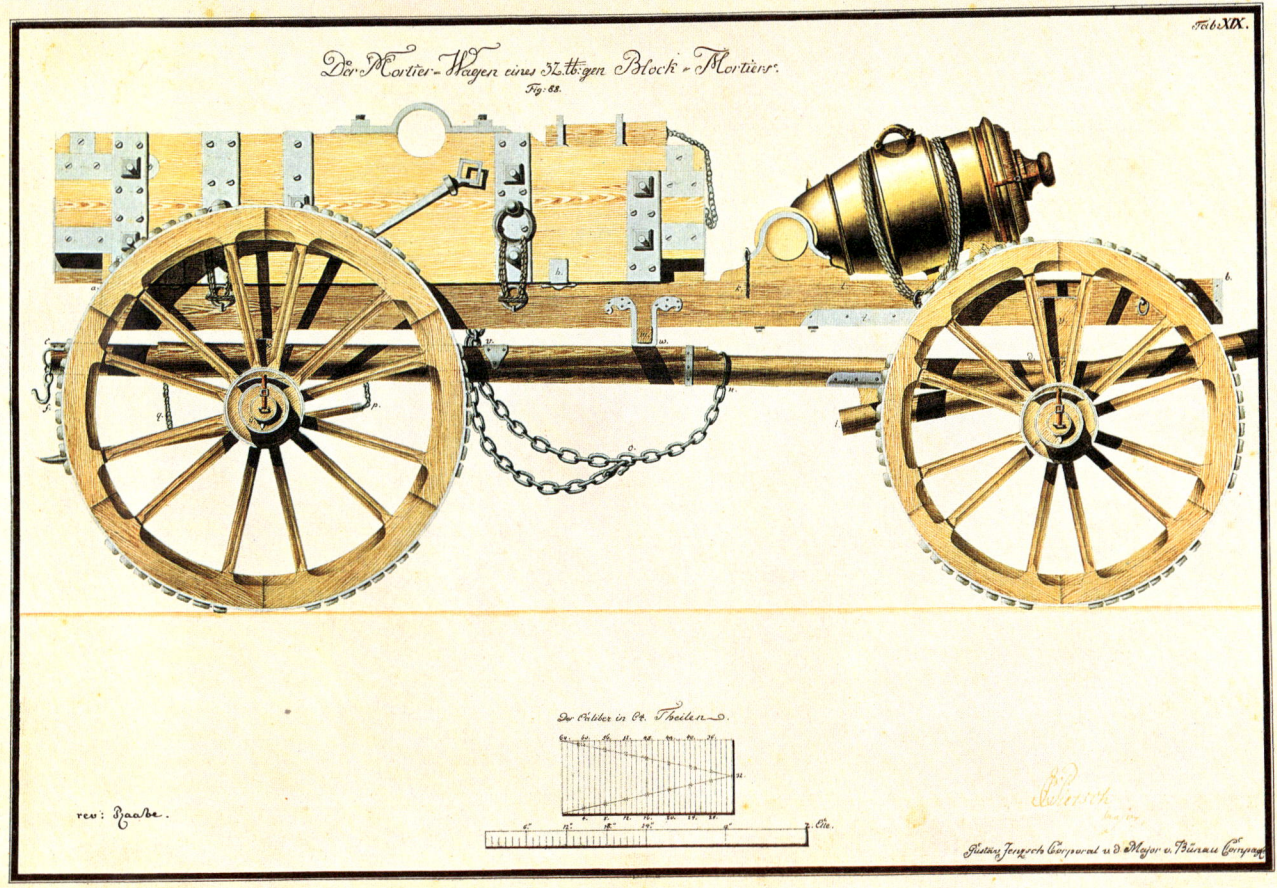

Der Caliber in 64 Theilen.

Gustav Jentzsch Corporal u. d. Major v. Bünau Comp.

For distance measurement by triangulation, the artillery officer first laid down a "base-line" with range-poles, as illustrated on the left. From each of the lower ends he then made a sighting on the target, as shown on the right.

cumbersome to be of much practical use in the field. The Germans had the idea of casting guns that were shorter and less thick than cannon, lighter and more mobile than perriers, initially loaded with grenades or shells filled with gunpowder. This explains why they were originally called grenade guns.

It was claimed that the length of a howitzer should be such that the soldier who inserted the charge could place it in the chamber by hand, without using any implement. Whereas the length of a mortar did not exceed two calibres, that of a howitzer was two or three calibres. But no records of any systematic experiments are available from which the length and weight of howitzers or mortars can be accurately ascertained.

In France, the denominations of mortars and howitzers indicate muzzle diameters; in Prussia, they refer to approximately half the weight of the projectile while, in Denmark, they indicate the entire weight of the projectile. It can be seen that, unlike cannon, no general rule had been laid down as to the definition of these pieces.

The gun-carriage used for the howitzer was almost identical in every respect to that used for cannon, the main difference being in the laying mechanism, which, of course, had to allow for a greater depression

of the breech to give the corresponding muzzle elevation for firing at an angle larger than 45°, or upper register fire. The breech of the howitzer rested on a quoin which moved forward or backwards in a groove in the laying block. A horizontal screw moved the quoin forwards or backwards, thereby lifting or lowering the rear end of the barrel.

Each mortar or howitzer was composed of a breech reinforce and a second chase reinforce. The part containing the powder was called the chamber and this could be cylindrical, spherical or cone-shaped. As with the cannon, the howitzer and mortar were positioned on the gun-carriage by means of trunnions. These, which not long since had been placed halfway along the length of the barrel with the mortar, were now at its base.

Bombs

The Dutch were the first to fire bombs from mortars. Francis Malthus introduced this practice to France and the French army used it for the first time in 1634 at the siege of Lamote in Lorraine. A bomb was a hollow sphere, provided with two handles and a hole called the vent. This vent was used to fill the projectile with gunpowder and to receive the wooden or metal fuze by which it was ignited. During the Siege of Gibraltar in

Left, the officer draws up the measurements of distances and angles he has just taken on a sketch-map. According to the scale, he can then calculate the distance from the target to one of the two extremities of the base-line, which indicated the gun positions.

1779, mortar shells with short fuzes were fired from the 24-pounder guns, at the suggestion of Captain Mercier, and this method burst the shells over the heads of the Spanish working parties. The fuzes used were of ribbed wood, with a powder core. Each rib marked a period of time, and the fuze could be cut at the desired rib.

Loading and Firing

The required quantity of gunpowder needed for the desired range was placed in the bottom of the mortar chamber, at the base of the bore. A wad, generally made of wood, was placed on top of the powder. It was so designed as to fill up any remaining space round the chamber. Over this

Obviously, the range of the weapons varied widely. The mortar had a much smaller range than the cannon, of which the heaviest had a striking distance of some 2,000 yds. Little was yet known about trajectory or ballistics.

tampion, a clump of turf was pressed down and moulded into a shape similar to the base of the mortar. Next, the bomb was inserted, fuze uppermost, and the bomb covered with earth, leaving only the fuse exposed. The mortar was now loaded.

A piece of quick-match was placed in the mortar's vent and attached with wax. Next a priming of fine gunpowder was poured in. With his port-fire, the bombardier first lit the fuze and then the quick-match in the vent. This meant it was ignited twice for one firing. It was only later that bombs were fired with one ignition, by lighting the fuze facing the muzzle, and fitted with a quick-match, with the charge. Numerous and often fruitless experiments were made before a really safe and reliable procedure was found to fire with only one lighting.

To attain the desired effect, the bomb had to explode after a certain time in flight, calculated according to the length of the trajectory. The bombardier knew that a certain five-inch fuze would burn in twenty seconds so he could shorten it to three inches if he calculated, on the basis of the charge and the barrel's elevation, that the fuze should ignite the charge after only twelve seconds. In 1674, when watches had only recently been invented and were still very rare, the great problem was to calculate the time needed for the fuze to turn down. Various devices were used, among which one which gained great acceptance from the pious was the repetition of the Apostle's Creed. The repetition of the doggerel: "If I weren't a fool, I shouldn't be here; I've left my wife, and all that's dear." satisfied the more ribald. Another method used was to attach a lead ball to a line c. 37 ins in length, and to fasten the other end to a stake planted vertically in the ground. The ball was then pulled approximately four or five ins out of the vertical and allowed to swing to and fro, each oscillation counting as one second. Using this rudimentary "metronome" and a simple rule of three, an officer or a well-trained bombardier could time the explosion of his bombs with punctuality and regularity.

At the end of the eighteenth century, guns no longer burst; they simply wore out. Their mobility had been considerably increased and, according to Charles Ailleret, they were comparable in this respect to the guns in use just before the mechanization of the First World War.

The projection of bombs paved the way for the shells fired by howitzers and cannon in succeeding centuries. They also inspired the subsequent invention of explosive shells of all varieties. The Siege of Gibraltar showed that the British had no really effective means of bursting shells, and it was not until Lt Henry Shrapnel came forward with his "spherical case" in 1784 that the problem was solved. His invention consisted of a hollow shell, filled with spherical shot, and containing a fuze and a small charge of powder. The fuze was calculated to fire the charge just above the target, and the charge was just sufficient to break the shell, letting the small shot continue in the shell's original line of flight slightly faster. The small charge ensured the shot were not dispersed. Shrapnel's invention, however, was not adopted by the British Army until the Napoleonic wars.

II. DEPLOYMENT OF ARTILLERY

A. ARTILLERY ON THE MARCH

From a manuscript written by Vasselieu at the beginning of the seventeenth century, it can be assumed that an army on the march formed a single column. Divided into three parts, it was formed by the advance guard, the main body and the rearguard, each of these sections comprising cavalry, infantry, artillery and transport. Owing to the build-up of forces, it soon became apparent that this disposition was ill-advised. The transport vehicles encumbered the columns on the march and the infantry, always slow to take up its combat positions, was slowed down even more.

In the eighteenth century, most towns boasted their own artillery. This engraving shows a Nuremberg battery in 1733. In the foreground, an officer (1) with sappers and carpenters (2 and 3) is followed by a second officer (4) in command of a wind & brass band and two quartermaster-sergeants (5 and 6). The third officer (7) rides ahead of his guards (8), a group of fife-players and drummers (9), the standard (10) and the first half of the gunners (9). A fourth officer (11) leads the gunners (12), four 4-pounders, eleven 6-pounders and one heavy gun drawn by six horses (13 and 14). The train (15) and the second half of the artillerymen (16) follow while an officer and his guards bring up the rear. The regiment is making its way to the town for a parade.

It was then decided to arrange for the movement of troops in separate formations, the artillery marching in the rear and just ahead of the transport columns. The shortcomings of this new method were soon evident. The columns of cavalry and infantry were strung out over too long a distance. In the event of a skirmish with the enemy, the artillery, far behind and moving comparatively slowly, arrived too late at its position in the line or on the flanks of the army.

The column order was changed again. The cavalry, infantry and the so-called active part of the artillery advanced separately in parallel lines, the artillery in the middle. Further back but also in the centre came the rest of the inactive part of the artillery and its transport.

Finally, during the eighteenth century, each tactical column, including the advance and rear guard, was provided with a detachment of artillery to cope with enemy attacks The distinction between active and inactive artillery was maintained and, on the march, the two groups were entirely separate.

The following extracts from the French regulation of 1778 provide a more detailed description:

"The artillery park of the army will be split up into six divisions. The first will comprise six 12-pounders and six 8-pounders; it will be called the artillery division of the advance guard and will be detailed to march with the battalions of grenadiers and riflemen whenever they assemble...

"... Four other divisions of artillery will be formed, each having a quarter of the pieces of ordnance and ammunition carriages which compose the army's artillery. Each of these divisions will be attached to a division of infantry and parked with it in the place assigned to it, either at the head of or between the two lines. It will march behind the infantry division, in the same column. It will be called the first, second, third or fourth artillery division, according to the number of the infantry division to which it will be attached.

"The heavy pieces will form the sixth artillery division; it will be shown where to take up position and assigned to column it must march with...

"... The heavy artillery will always march with the column which is the best and will follow the small and large train of this column".

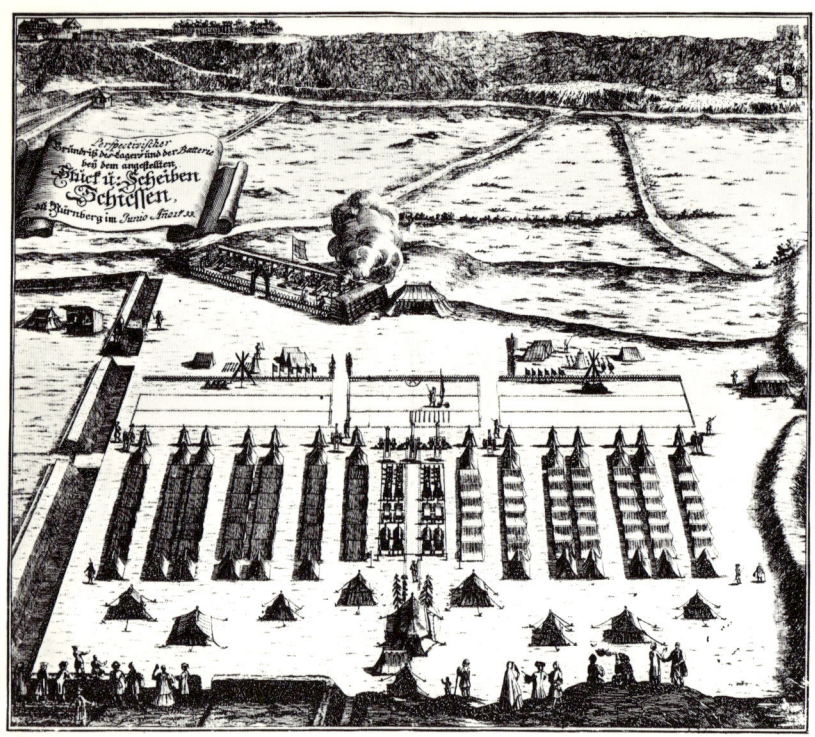

After the parade, the regiment returns to its quarters and the guns are moved into their emplacements. The cannon are placed in the firing position behind a defence system and a target set up a few hundred yards away. The presence of this target shows that the artilleryman always fired directly, at point-blank range.

In a town such as Nuremberg, the artillery review provided an ▷ *opportunity to inspire an esprit de corps in the militia and to demonstrate in public the excellent state of readiness of those who might be called upon in its defence. The scene illustrated opposite displays the exemplary order and discipline reigning in Nuremberg as the Burgomaster arrives in his carriage to witness the shooting. In some towns, the popular flavour of the occasion sometimes obscured the strictly military character of such a review.*

DEPLOYMENT OF ARTILLERY IN THE FIELD

Towards the end of the seventeenth century, artillery managed to arrive on the battle-field just as the infantry was forming into lines. While these troops, always very slow, were manoeuvring and lining up in their determined positions, the officers made a reconnaissance of the terrain and chose the positions to be taken up by the artillery brigades. In fact, this choice was made in a very simple way: all the guns were placed in single line in front of the spearhead of the army. The only variable factor was the number of guns allocated to each brigade. During the reconnaissance, the officers and troops took possession of the guns, material and transport and organized themselves into brigades.

At a signal, the organized brigades marched past the Artillery General who allocated battle stations. Each brigade, in its turn, was inspected by its commanding officer and made its way into position,

In the distance, infantrymen march past in parade order before taking up their battle stations. In the foreground, artillerymen are hoisting a mortar barrel on to its bed, while behind them a gun-carriage stands ready for its cannon. This engraving, albeit somewhat academic, shows that gunners had to be prepared for all the unforeseen situations arising from mere accidents or the hazards of war.

passing behind the troops which were already drawn up. Arriving on a level with its appointed position, it passed through the gaps in the troops and took up its station a hundred yds ahead of the line, and was supplied with twenty-five to thirty rounds per gun.

As soon as the guns had been unlimbered, the horses and limbers were taken back between the two lines and stationed next to the ammunition wagons, thus forming a brigade park in the rear. The battle began with fire from the artillery, the infantry remaining stationary behind the guns until the signal for attack. "The artillery fire lasted a long time", wrote Brunet in his *General History of Artillery*, "powerfully affecting the morale of the troops and their commanders, so that armies often withdrew without bringing their troops into action".

This cannon has just been brought into position, the powder and cannon balls unloaded, and the gunners are enjoying a moment's respite. The carriage and limber are covered with a tarpaulin and are about to be pulled back to the rear. Meanwhile, the infantry forms up into battle order.

While the battle rages in the distance, these gunners, sheltered behind a convenient hillock, hold a private council of war. One of their horses has been fatally wounded. Luckily, reinforcements arrive to defend the position or manhandle the gun to another site.

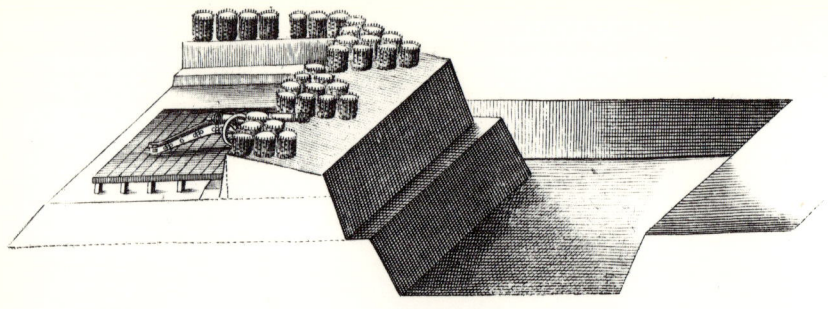
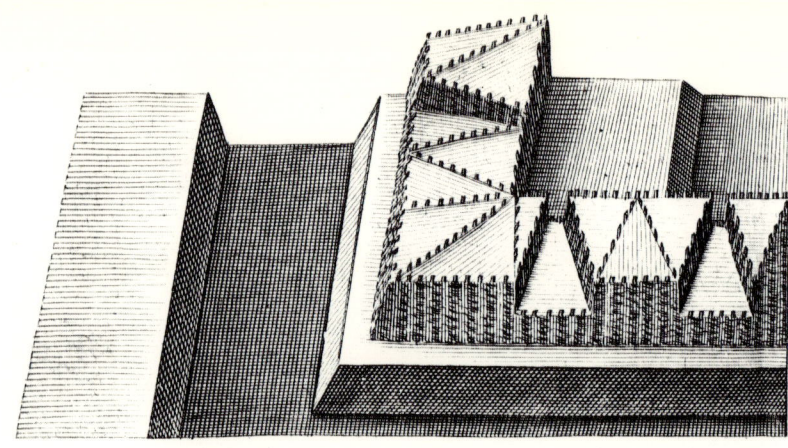

During a siege, the cannon, howitzers or mortars were set up behind a moat and a breastwork. They were protected by gabions, as seen on the left, or by earth bound together by faggots, as on the right. Platforms made it easier to bring the guns into action.

Towards the middle of the eighteenth century, two facts became obvious. The superiority in arms and the large distances to be defended made it imperative to extend the battle lines and, except in those rare cases where artillery accompanied the assault forces in order to clear a way through the enemy lines, the defensive was considered the most favourable position.

Wider use began to be made of entrenchments and, whenever weather permitted, armies maintained permanent lines of trenches, manned by the entire artillery. The troops remained to the rear. Brunet commented: "In such a position, the armies, as if imprisoned, degenerated into a continual state of inertia and cannonades became the only means of action. So on a number of occasions, armies could be seen advancing towards each other through the entrenchments protected by the battery."

This evolution in the use of artillery provoked comments from several military writers. In his *Essai Général de Tactique*, published in London during the year 1772, the Count Jacques de Guibert wrote: "The cannon, considered on the basis of its individual effectiveness and aimed at an isolated object of small area, is a weapon little to be feared, if at all. But it is not used in this manner in combat. At such times, it is not a question of a single point but of entire lines, of massed formations of troops. Then, if artillery is to be used, large batteries are formed to pound, not specific points, but wide areas and cover gaps in the target lines. Ricochet fire is employed, lines of battle are extended ... artillery does not set itself the small objective of disabling one cannon or killing a few men but the large,

decisive objective of covering and sweeping with gun-fire the ground occupied by the enemy and through which he hopes to advance."

In his work entitled *On the Use of the New Artillery*, published in Metz in 1778, Jean du Teil defined the mission of the artillery officer in open warfare. This consisted of examining all openings and irregularities of the terrain which could screen the enemy or enable him to advance under cover of fire, to simultaneously reconnoitre new emplacements in all possible directions and seek the means of reaching them without being observed.

The reason for this alteration in the deployment of artillery was that troops had adopted the shallow battle formation, the infantry in three ranks, the cavalry in two. Ineffectual because of its lack of accuracy, the artillery sought to compensate for its shortcomings by the use of ricochet and cross-fire. It no longer took up its position in front of the troops but in the spaces between them.

Guibert stipulated: "Apart from the mutual protection which the batteries should try to give each other, they must be made strong. Then they can achieve decisive results, open breaches, pave the way for victory .... It does not necessarily follow, from the foregoing principle, that too large a complement of artillery should be combined in one and the same battery. That would create another pitfall, of laying themselves open to enemy fire. It is simply a matter of training several batteries, a short distance apart, on to the same object..." In more modern terms, this amounted to concentration of fire.

Since it did not form an independent branch of the army and did not follow any systematically planned procedure, the role played by the artillery during battle is difficult to describe in any greater detail. Artillery prepared the attack for the infantry lines or, conversely, supported the line when it was in difficulties. It was under Bonaparte that artillery ceased to be an adjunct of the infantry.

With a little imagination, it is easy to picture a battle-field of those days, where some thousand guns thundered out in the midst of hundreds of thousands of men. The armies were drawn up

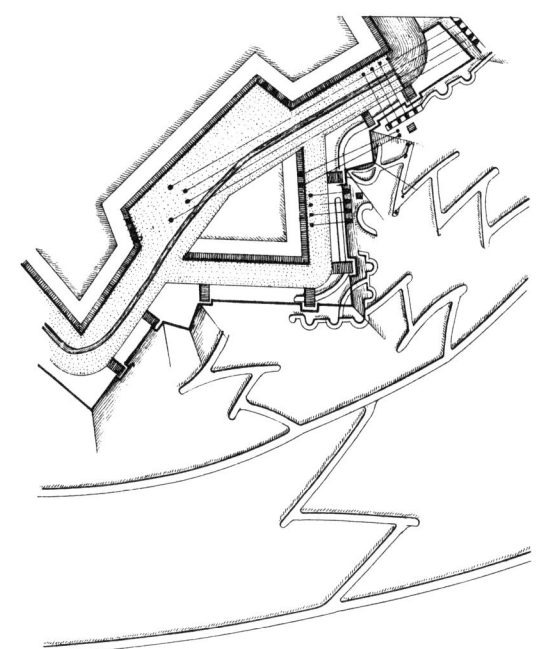

The siege of a town was often a long and arduous undertaking. Through open or covered trenches, sappers approached within firing range of the enemy fortifications to prepare the battery emplacements according to a general plan, as shown in the sketch right. Each battery was arranged as in the model below. Behind the embrasures in the parapet, the guns were placed in position. In a trench further back were small stores of powder and cannon-balls and, at the rear, the main powder reserve.

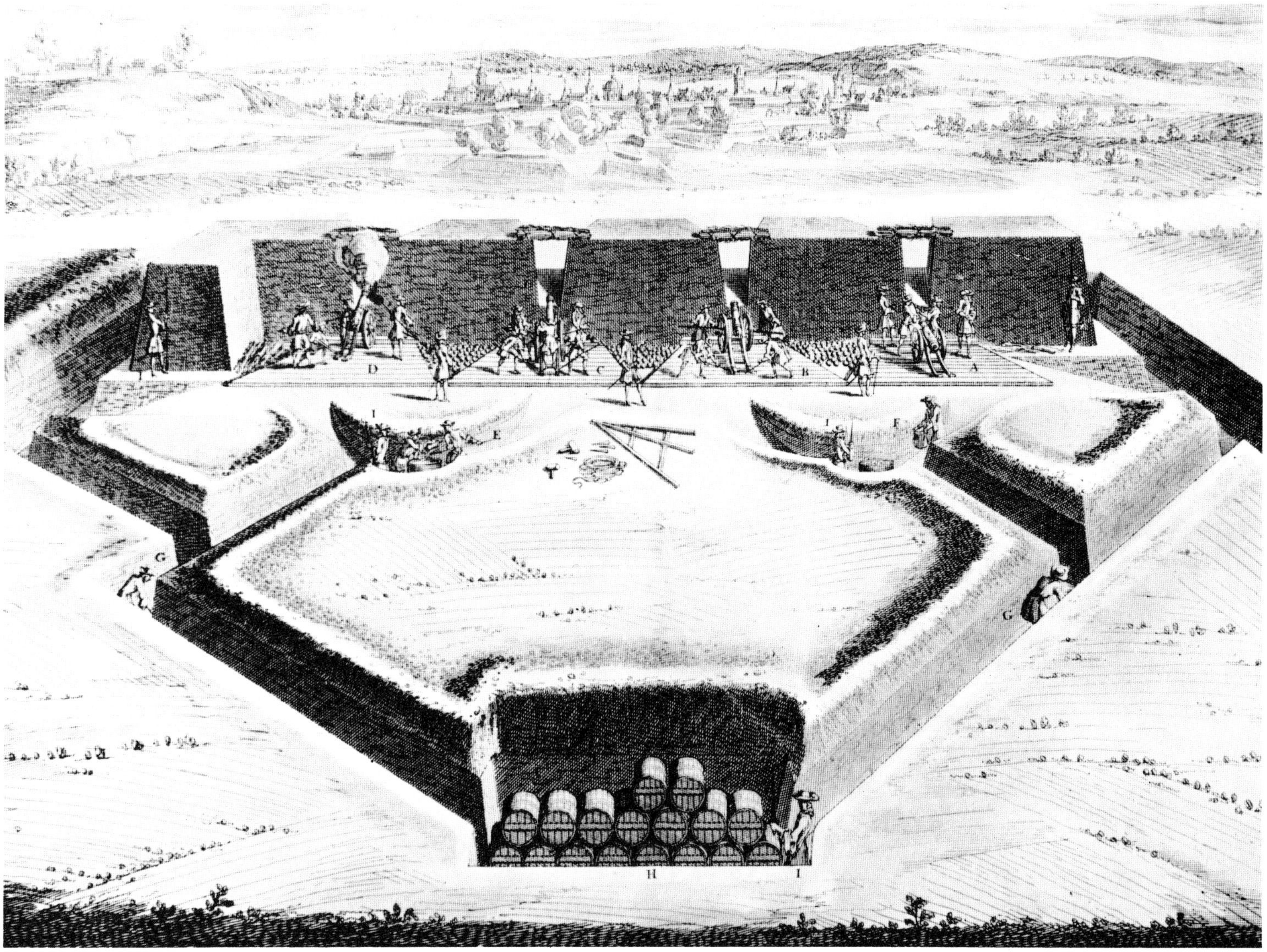

facing each other, protected by numerous batteries. Behind the artillery stretched the long lines of foot-soldiers with arms at the ready. In the spaces between the brigades and the divisions were the batteries attached to these corps, their escort throughout the engagement. On the wings, the squadrons of cavalry were massed while, further to the rear, a few small reserves of artillery were held in readiness to take up positions should the development of the battle require it. Finally, at the rear of the entire army, was stationed the main artillery reserve which would not go into action until such time as the commander-in-chief used it, either for defence or to assure victory.

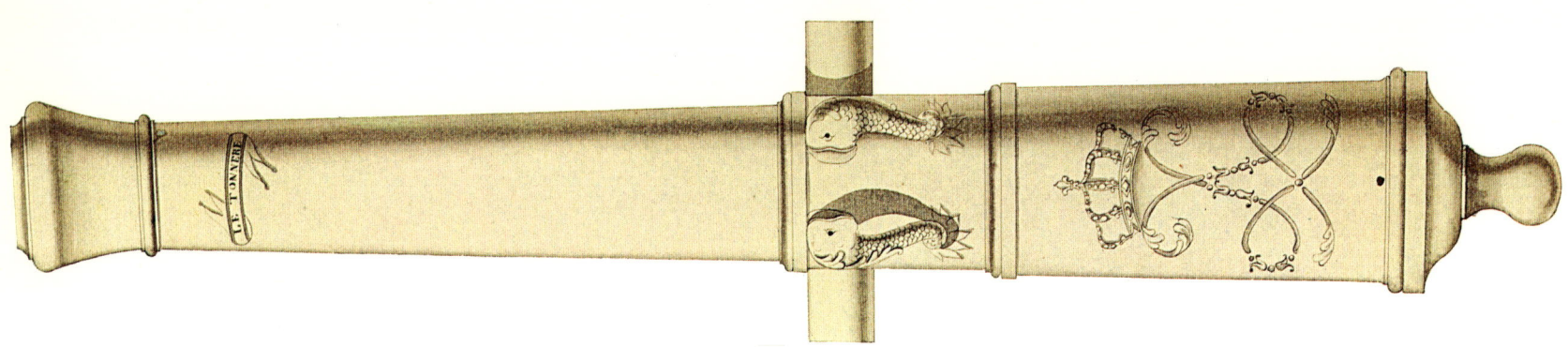

The 24-pounder siege cannon, above, was the heaviest piece in the Gribeauval system of 1765. The diameter of its projectile was approximately 6 ins, the gun weighed 2.7 tons, was supplied with 1,000 rounds of ammunition and had a range of about 1,800 yds. The plan and elevation of the gun-carriage are shown below on the left and its limber on the right. Shaft draught is retained and although the limber no longer incorporates a tool-chest, an ammunition box could be placed on the gun-carriage, near the transom of the trail.

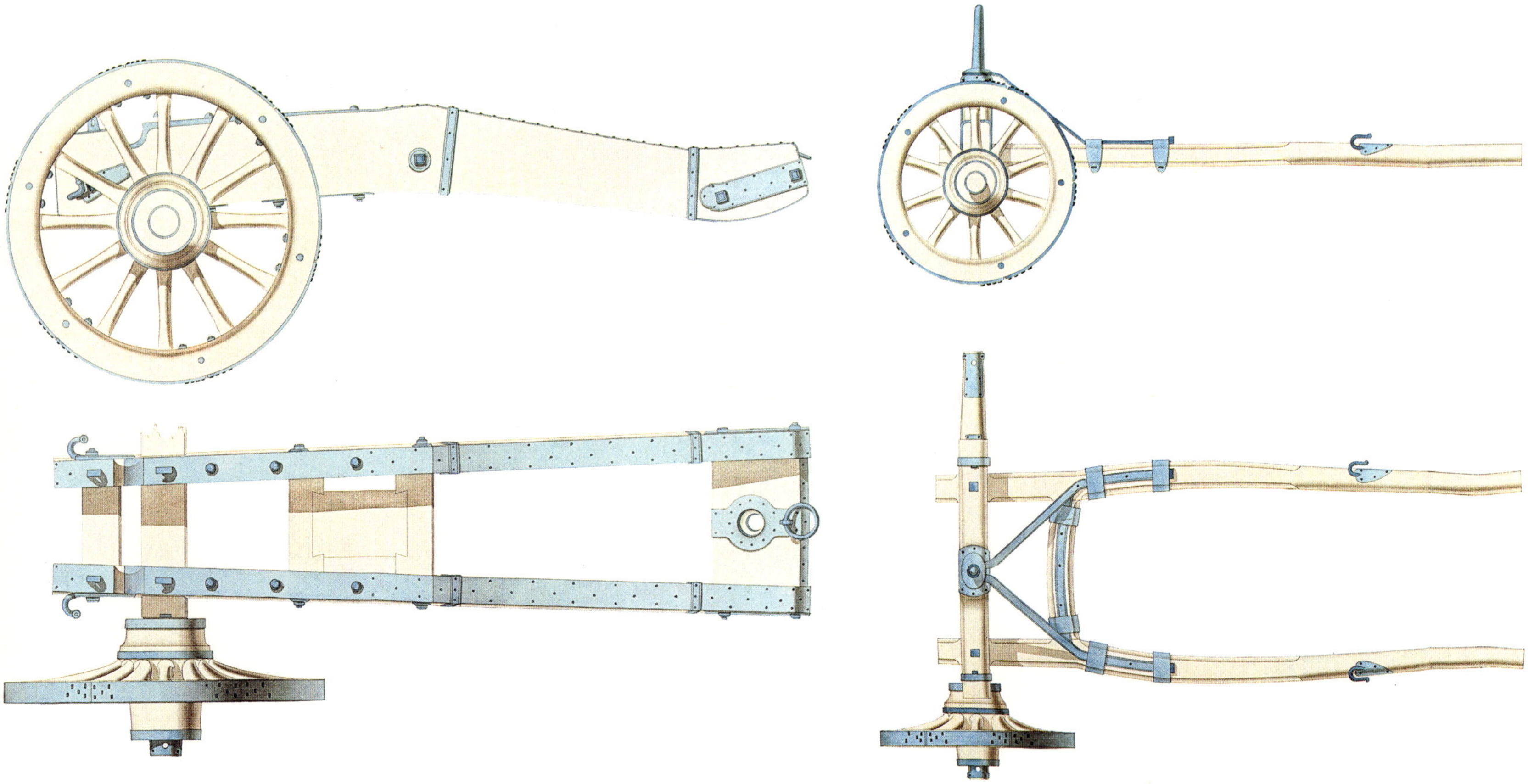

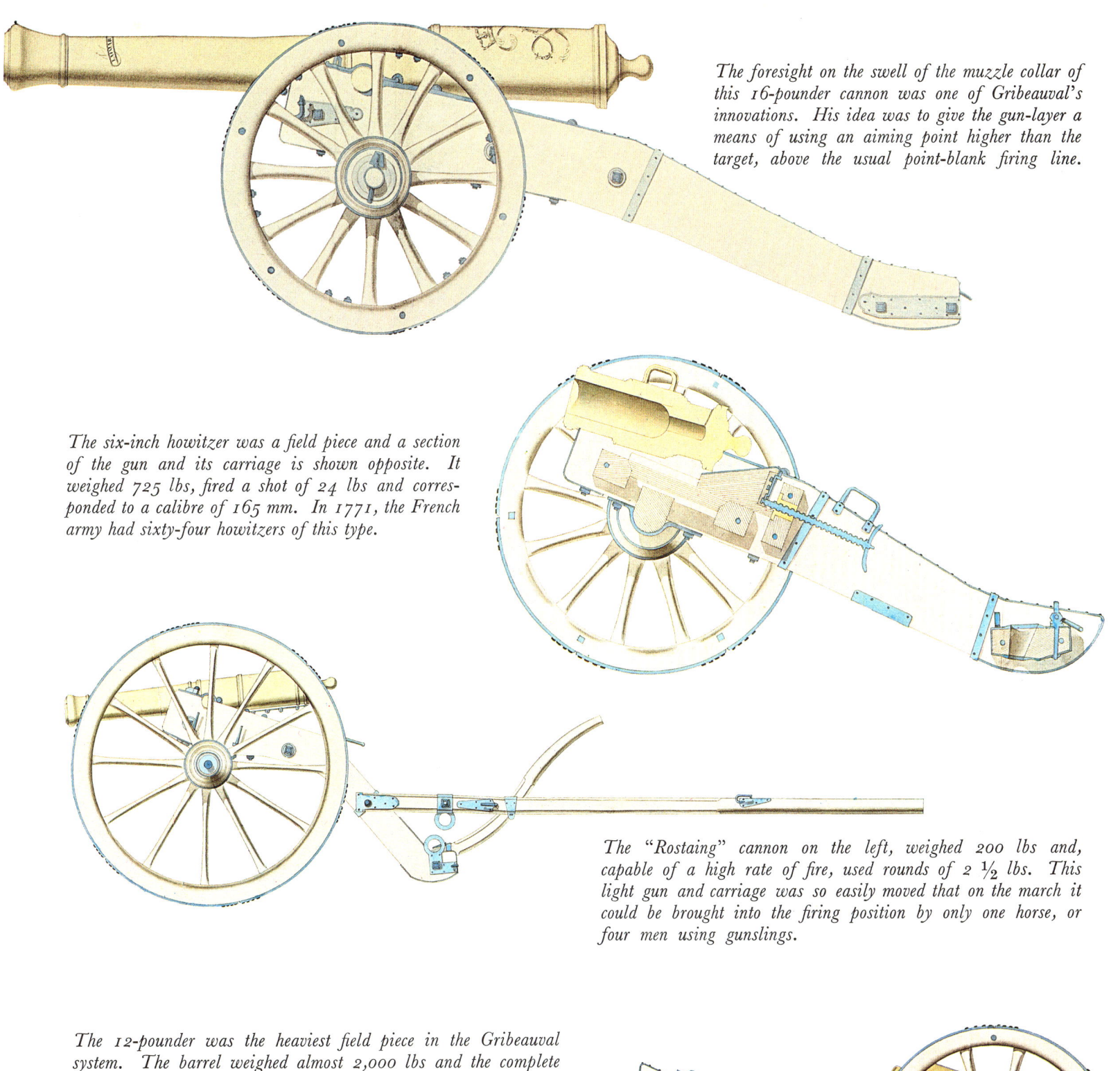

The foresight on the swell of the muzzle collar of this 16-pounder cannon was one of Gribeauval's innovations. His idea was to give the gun-layer a means of using an aiming point higher than the target, above the usual point-blank firing line.

The six-inch howitzer was a field piece and a section of the gun and its carriage is shown opposite. It weighed 725 lbs, fired a shot of 24 lbs and corresponded to a calibre of 165 mm. In 1771, the French army had sixty-four howitzers of this type.

The "Rostaing" cannon on the left, weighed 200 lbs and, capable of a high rate of fire, used rounds of 2 ½ lbs. This light gun and carriage was so easily moved that on the march it could be brought into the firing position by only one horse, or four men using gunslings.

The 12-pounder was the heaviest field piece in the Gribeauval system. The barrel weighed almost 2,000 lbs and the complete piece approximately two tons. Its cannon ball weighed 13 lbs. Pole draught has replaced shafts.

As has already been seen, the battle began with the first cannon shots. Gunners rarely used the maximum range of their pieces for, at an elevation of 8° or below, the large calibres had a range of 2,000 yds and the 4-pounders approximately 1,600 yds with an elevation of 6°. During action, the 12-pounders and howitzers fired shots reaching between 700 and 1,000 yds, the 8-pounders making 850 yds and the 4-pounders as much as 800 yds. Between 600 and 400 yds, all guns fired large ball-cartridges and, from 400 yds, small ball-cartridges. Ammunition fired from the field guns could penetrate several ranks of soldiers, smash through a roof or fell a tree. Nevertheless, it is difficult to accurately assess their power of penetration. All that is known is that during siege warfare, shots fired from a distance of 300 yds penetrated to a depth of between 1-4 yds, depending on the nature of the soil.

Against troops, only ricochet fire, shells or case-shot caused appreciable losses. A solid shot which does not ricochet is not an especially dangerous missile whereas ricochet fire extends the danger zone to 40, 60 or sometimes 200 yds behind its first point of impact, and was difficult to avoid.

The 12-pounders and 8-pounders, at an elevation of 8°, fired at the leisurely rate of one or two rounds a minute, the 4-pounders at the rate of three shots per minute and the Swedish-type 4-pounders at as high a rate as eight or ten shots a minute. As for the 24-pounders and 16-pounders, they fired slowly because of the lengthy procedure involved in returning these mammoth weapons to the firing position. A spirit of rivalry and *esprit de corps* could engender great feats of skill or endurance on the part of the gunners, often compensating for various deficiencies in the equipment or even mitigating the disadvantages of bad ground conditions. However, the gunners were hampered by incidents that are hard to imagine nowadays. As each shot was fired, the wad was simultaneously ejected, scattering in blazing splinters which, depending on the region and the season, often started

These two guns, decorated with the badge of St Mark, were cast in Venice in the presence of Frederick IV, the King of Denmark and Norway. The culverin, left, was cast in 1709 by the Master Founder Joannis de Mazzaroli while the cannon dates from 1708. In Venice, even guns were works of art.

dangerous fires. Each shot was also accompanied by a cloud of dense smoke, forming a screen between the gunners and their targets, thus considerably slowing down the rate of fire and jeopardizing its accuracy. Another incident, not uncommon, was for some guns to become "unbushed". This meant that the hardened tube that lined the vent-hole to prevent it becoming eroded by the blast of abrasive grains of powder was blown out. Fitting a new bush, which had to be fed into the vent-hole from the bore, was a laborious task, for the bush had to be shaped on the spot to fit the individual gun, and the gun fired with reduced charges until it was properly seated in the vent. After the battle, the artillerymen spread out over the battle-field to salvage those rounds which could be used again.

D. SIEGE ARTILLERY

When the purpose of the military operation was to capture fortified positions, artillery played the leading part. Artillery, alone, could break down fortifications and defence works of all types.

While the army surrounded the fortified position and cut off its communications with the rest of the area, the artillery set up a vast system composed of heavy guns, tons of gunpowder and shot, wooden and iron equipment, great quantities of tackle, such as machines and tools, as well as a wide range of armaments and workshops of every kind.

The transport and conveyance of this gigantic formation depended on the kind of lines of communication, water-ways in particular being preferred, and posed enormous problems.

Having arrived in the vicinity of the fortified position, the siege artillery stationed itself out of sight and fire from the besieged town. The army officers assigned their troops working positions, depots, parking sites and camps. Either by means of material brought with them or by use of local resources from neighbouring forests or requisitioned in the surrounding towns and villages, a veritable town of wood and canvas gradually took shape, defended by the infantry and cavalry who prevented any sallies from the besieged town. Well supplied

with howitzers, the besieged could sometimes range on the camp and penetrate the cover of the besieging army. Whenever possible, the besiegers pitched their camp beyond the anticipated range of the enemy artillery.

From this encampment, sappers dug trenches through which guns could be brought up under cover to their assigned positions. All the digging was carried out at night, the men throwing the excavated earth forward, in the direction of the beleaguered town, so as to form a protective embankment quickly.

Meanwhile, the officers in command of the batteries made the necessary arrangements and staked out the camp. At nightfall, the sappers began digging the battery emplacement, this time throwing the earth backwards, away from the fortified position, so as to form a ditch in front of the protective embankment. "I know", wrote Surirey de Saint-Rémy in this connection, "that the soldier is apprehensive when working in the open, but also a battery which is mounted in this way is much stronger and less men are lost in the course of the siege than in one where the earth is thrown ahead and where troops are necessarily obliged to fetch more earth from behind the exposed battery to replace that which has been removed in order to raise the level of the ground occupied by the gun platforms which is too low to receive the guns, the latter sinking in because of the lack of firmness of the ground beneath them."

Day after day and night after night, work progressed and the position of each battery was gradually consolidated. Once completed, a battery of four guns with their gun crews were dug in behind a breast-work thirty-five to forty yds long, six ft high and six yds across, the walls being made of faggot and earth-filled gabions to bind the construction together. The guns were placed at intervals of approximately seven yds, each opposite a wedge-shaped embrasure, mounted on a platform of thick boards raised at the rear end to check the recoil movement. Trenches, protected by embankments, led to the powder magazines and ammunition stores and, further to the rear, into the camp itself.

Winding straw rope around a wooden core resting on two trestles was the first stage in moulding a cannon. This core was timber cut to the required length and calibre of the barrel.

Once the core was covered, the moulder applied layers of clay, which was then carefully dried to prevent cracks.

Trunnions and the ornamental details for the decoration of the gun were prepared in separate moulds and then applied to the model in the required position.

Siege artillery consisted primarily of heavy guns. Surirey de Saint-Rémy quotes the following figures when describing the artillery train for a siege at the end of the seventeenth century: fifty 33-pounder and 24-pounder cannon, ten 16-pounder cannon, ten 12-pounder cannon, twenty 8-pounders, twenty 4-pounders and forty mortars. For a siege lasting thirty days, this artillery warranted a supply of 12,000 cannon balls of 33 lbs and 19,000 of 24 lbs. As much as 36,000 fathoms, or 72,000 yds of quick-match would be required while the gunpowder alone would weigh approximately 720,000 lbs.

The heavy guns were used almost exclusively for breaching, or to bring down a wall or bastion to pave the way for an attack. The 33-pounder and 24-pounder batteries either concentrated their fire or spaced it out with the object of shaking the fortifications. A battery consisting of ten to twelve 24-pounders, each firing between twenty and one hundred shots per day, could make a breach in twelve to fifteen days. By increasing the number of guns employed to batter the same objective, the breach would obviously be made more quickly. The 16-pounder and 12-pounder batteries acted as watch-dogs, firing at covered communication paths, at cavalry in action or, if necessary, at the breastworks if the batteries were threatened.

The lighter 4-pounder and 8-pounder guns defended the approaches to the manned or communication trenches, and sunken roads.

At night, if it was intended to set fire to the fortification, red-hot salvoes were fired from 8-pounders or howitzers. Mortars and howitzers were able to reach places that were inaccessible to the cannon. Naturally, the gunners in the besieged town would use their heavy pieces in an attempt to neutralize their enemy's batteries by demolishing the defensive positions, killing the gunners and destroying the guns and stores.

In all siege warfare, artillery formed the backbone of the military undertaking. Everything else revolved around it and it was during this period that the troops became fully aware of the interdependence of the various branches of the army. Artillery was powerless without the help and protection of the infantry. The infantry and cavalry, so long accustomed to taking the leading part in any battle, simultaneously took stock of their own limitations and of their reliance on the strength of the artillery. An element of improvisation also played a part in sieges, particularly on the part of the defenders. In 1781, the Spanish and French combined to try to recapture Gibraltar from the British. After a bombardment lasting eighteen months, the grand stroke was to bombard the Rock from specially-constructed floating batteries. The fortress pieces on the Rock could not be sufficiently depressed to deal with this menace, so Lt Koehler invented a depression carriage which solved the problem, and, with the use of red-hot shot, enabled the Britsh to destroy the floating batteries and a large part of Spanish fleet on 13 September, 1782.

CANNON PRODUCTION

Until the beginning of the eighteenth century, all cannon were hollow-cast, either by the muzzle or the breech. The latter method had been perfected by the Keller brothers, Hans-Jakob and Johann-Balthazar, who had been invited by the King of France to set up the foundry of Douai. Once the cannon barrel had been cast and withdrawn from the mould, it was placed vertically, muzzle downwards, in the drilling machine. The tang of the borer also served as the spindle of a treadmill powered by horses. The weight of the barrel progressively pressed it down on to the boring machine. This method did not prove entirely satisfactory. In 1704, Jean Maritz I, a Swiss from Burgdorf, successfully made solid castings with horizontal boring for the barrels. With this system, the cannon itself turned about a fixed drill.

This achievement probably represents the most important development in the manufacture of cannon during the eighteenth century. Jean Maritz I first worked in Burgdorf then, from 1723, in Geneva and, from 1734, in Lyons bearing the title of *Commissaire des fontes de l'artillerie*. He died in Geneva in 1743. The Maritz family proved themselves to be except-

Next, the cope was made by applying three thin layers of clay to the model. Each successive layer was dried by light firing and the cope was then reinforced with iron bars and hoops.

Now finished and reinforced, the cope was dried out. The core and coiled straw sleeve were carefully taken out and an exact impression of the original model was left in the shape of a mould which completed the process.

In preparation for the delicate casting process, the moulds were placed vertically in casting pits, in front of the furnace. The molten metal was poured through sow channels, filling the remaining empty space inside the copes.

When cooled, the whole unit was removed from the pit and the cope broken open with hammers to free the barrel.

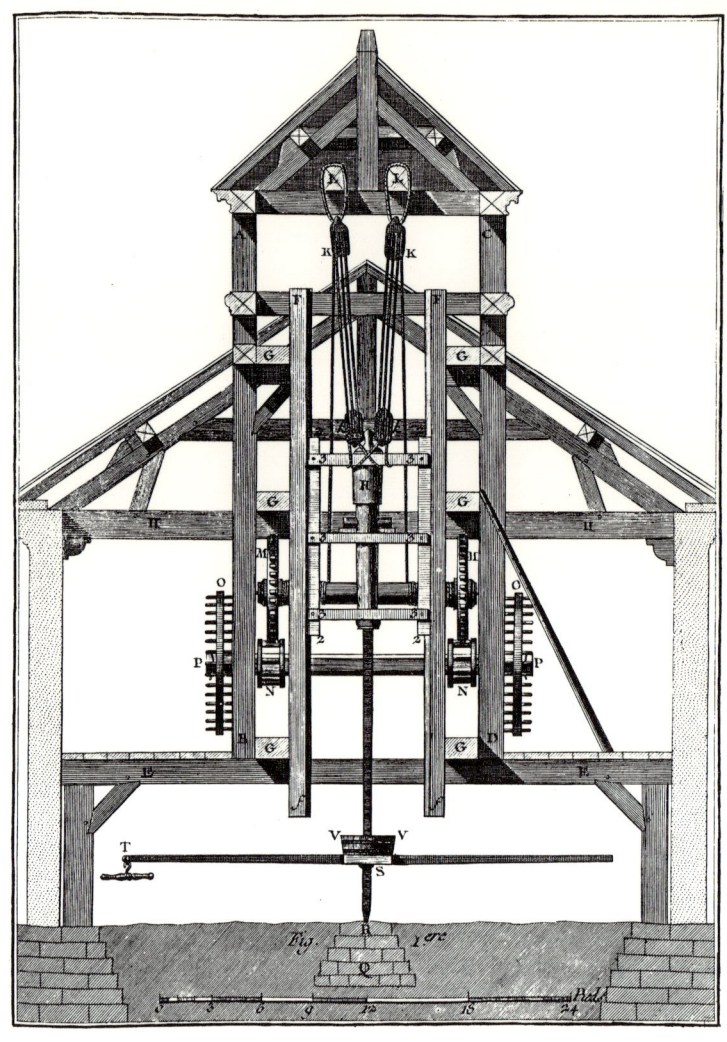

This drilling machine, mounted vertically, was used to bore out the solid casting to the required calibre.

ionally good founders and their services were much in demand by sovereigns and heads of republics.

Jean Maritz' second son, Jean Maritz II, continued his father's work in Lyons and subsequently in Strasburg, Douai, Rochefort and Ruelle. His services were greatly appreciated. In 1744, Louis XV granted him a pension of two thousand pounds and, in 1755, bestowed on him the title of *Inspector General of the Casting and Ironworks of the Artillery* with a salary of £15,000. Jean Maritz II also set up and reorganized foundries in Spain, at Barcelona and Seville, with the title of *Inspector General of Casting and Ironworks of the Artillery of Spain* with brigadier's rank. Catherine the Great of Russia, hearing his praises sung, tried to lure Jean Maritz II to St Petersburg but he resisted her blandishments.

In 1768, "in consideration of his services for thirty-four years", the King of France bestowed on him a pension of £12,000. He died in 1790, at the age of seventy-nine.

Samuel, the elder son of Jean Maritz I, succeeded his father as head of the foundry in Geneva. In 1748, he took over the management of the foundry at Berne and, ten years later, reorganized various foundries inside the territories of the German Empire. In 1786 he died, blind, in Geneva. His son, Jean, managed the Barcelona foundry and, later, the foundry at the Hague until his death in 1807. This family business had contributed greatly to the development of the manufacture of pieces of artillery.

Moulding

The first stage in the moulding process consisted of making a solid pattern. Straw matting was wound around a template, a board of appropriate size to the calibre resting on two stands. The whole was then covered with paper so as to isolate the straw from the clay, which the moulders subsequently applied in successive layers to produce the exact shape of the desired cannon. Trunnions, lifting handles, etc, were applied to the model with great precision. During this time, a wood or peat fire was constantly tended and kept burning under the template in order to dry the clay more quickly. The entire model was then covered with two thin layers of tanner's cinders, mixed with water, to prevent the model adhering too firmly to the cope that was to be built around it.

The next stage was to construct the cope, an outer covering or shell made from layer upon layer of clay, each successive layer being fire-dried. The cope was then reinforced and bound together with iron bars and hoops.

The third stage was the tricky procedure of first removing the template from the cope, then the straw matting. The cope, now hollow inside, was rebaked by laying it in a foundry pit into which fire-brands or blazing faggots were thrown until such time as the cope was thoroughly set. All remains of the original mould could then be removed, preserving only the cope whose inner surface retained the shape and whatever ornamentations were on the mould.

In the traditional method, a long piece of iron, the *core*, of the same dimensions as the gun-bore, was then placed exactly in the middle of the space thus cleared. This core was held accurately in position at the breech by crossed iron bars and at the muzzle by a special framework. The empty space between the cope and core received the molten metal and corresponded to the thickness of the gun walls. A coating applied to the entire surface of the core prevented it from sticking to the cast metal and facilitated its removal once the gun was cast. The moulding process was now completed.

From Jean Maritz' time, and by means of an entirely new process, cannon were also cast solid and afterwards drilled to produce the bore.

Here, skilled engravers take over the newly cast gun to give it the finishing touches. They polish off the burrs, clean up the chasing and inscriptions, leaving it with all the grace of a royal arm. Finally, the gun was weighed. In this illustration, a steelyard is used, hung from sheer legs. In this way, the Master Founder checked the exact weight of each piece and, therefore, the density of metal, before delivery.

Casting

The cope was placed in a casting pit, muzzle or breech uppermost, depending on the method favoured by the Master Founder, facing a furnace where the alloy was brought to its melting point.

The founder then poured the molten metal from the furnace into the cope through small channels. The liquid metal filled the remaining empty space inside the cope and the gun thus took shape. This was a delicate operation as the founder had to ensure that the gun metal was of uniform consistency throughout. Once the casting was finished, the piece was left to cool.

Boring

Once cooled, the cannon was removed from the pit in its unformed state and stripped of its clay casing and metal burrs. The piece was then bored by cleaning and enlarging the bore so that it corresponded as closely as possible the calibre.

Each type of boring, either horizontal or vertical, could be carried out in two ways, either by turning the fixed drill and moving the cannon forward, or by turning the cannon and moving the drill forward. At the end of the eighteenth century, the usual practice, at any rate in France, was to bore the solid cannon horizontally with a fixed drill. The machine was driven by horses, by water or even by steam-engine. The vent was also bored and a bush screwed into the breech. This was a cylinder of beaten copper bored with the actual priming-hole or touch-hole. In practice, when this hole became too open at the mouth, it was replaced. Each bush could be used for between three and six thousand shots before having to be changed. On guns of large calibre, bushes were replaced as many as three times.

Gun Trials

Tests consisted of examining the gun, weighing it, ensuring it had no defects or malformation and firing a number of test shots, usually five. When the gun had fired the specified number of shots, it was suspended vertically, breech-knob downwards. The vent was sealed and the bore filled with water which was left for at least eight hours. After this, an inspection was made to ensure that the water had not leaked out, especially around the lifting handles and the priming-hole section. Only then the flawless guns entered the service of the King. The rejects were broken up and returned to the foundry, where they were melted down to provide a fresh supply of alloy for the casting of other guns.

THE NAPOLEONIC WARS
1783 - 1815

Towards the end of the year 1790, the last army of the French Monarchy was swept into the revolutionary storm which was to shake France to its foundations and unsettle all Europe. With flags flying, but without many of its former officers, this army brought its glorious history, its might and its devotion to duty into the service of the new regime.

The artillery was in fine fettle. Its organization was adapted to contemporary needs, its equipment excellent, its training methods first-rate, its officers able and well-informed. Moreover, it benefited from intensive theoretical studies carried out by the military tacticians.

On the initiative of Lacombe Saint-Michel, Gomer and Rostaing, all artillery officers as well as parliamentarians, and with the support of Narbonne and Lafayette, the French artillery which Guibert had considered as a "useful adjunct" rapidly developed into a military arm in itself. It was officially recognized as such on 29 October 1790 and administered as a corporate body by the "Artillery Committee". The artillery immediately adopted a democratic character and was given a "new look". The seven regiments, each divided into two battalions of ten companies, exchanged their name in favour of a number. All soldiers became "gunners" and the "seconds-in-command" would henceforth have the more important-sounding title of second officers. As in the regular army, a company of gunners with two guns was assigned to each battalion of National Guards who, although regular soldiers, were still called "Volunteers" according to a statute of 18 March 1792. One month later, war was declared between the two European nations having the greatest military might and the best equipped artillery.

The French army was fortunate in receiving the support of nine companies of horse artillery, whose formation had long been urged by Lafayette, Narbonne, d'Aboville, Sorbier and many others. At one time or another, all of them had said: "Well-used, horse artillery is the safest means of shielding the movements of indifferently trained troops". This fittingly described the French infantry, its ranks swamped by an unruly mob of pseudo-volunteers. On the other hand, discipline in the artillery was reasonably good, because of the equipment, which required good reflexes and accuracy to use well. Some officers had fled the country, such as Manson who now commanded the artillery of the Army of Condé, which fought on the counter-revolutionary side, but owing to the *esprit de corps* inherited from the old regime, enough competent officers still remained to maintain order and command the regiments. D'Anthouard, Dommartin, du Teil, Lariboisière, Sénarmont, Songis, Sorbier, Gassendi and Buonaparte, as well as many others such as Quartermaster-Sergeant Pichegru and the drummer Victor, were to put the rules of procedure and programmes advocated by the theoreticians of the eighteenth century into practice. They were also to make headway with the independence of the new arm, and the horse artillery, which was being formed into companies.

The corps of engineers still being at an embryonic stage, even the Rhine boatmen were combined with artillerymen into units on pontoons. Heavy artillery was used in all its varied functions: for sieges, in garrisons and on the field. A regulation in 1792 conferred the title of "Reserve Artillery" on those heavy batteries attached to the armies. Subsequently, by slow degrees, the "Volunteer Artillery", more a political than military expedient, together with mounted companies, gradually took over the best human and material elements of the heavy artillery, jeopardizing its traditional spirit and martial qualities. A regiment was a myth, its scattered units never coming together. It was impossible to find a division comprising eight cannon of the same calibre while the number of guns assigned to companies varied from two to twelve. As if this was not enough, the gunners' fusils were handed over to the infantry!

On the other hand, the mounted companies fared better. They were even provided with troopers, becoming increasingly rare. On 27 December 1793, the Committee of Public Safety decided on an increase in the number of horse batteries to one hundred. As there was a shortage of horses at the time, these were slow to materialize. In the meantime, gunners clambered on to the backs of the reserve horses or sat astride the padded lids of the limbers.

7 February 1794 was a memorable date in the annals of artillery in France. Horse artillery became an independent force and became known by the title of Light Artillery. It comprised nine regiments of six batteries, each theoretically consisting of four officers, seventy-two men, six 8-pounder cannon and two howitzers. These were usually employed in half-batteries and often in sections consisting of two cannon and one howitzer. Field artillery *par excellence*, it was highly esteemed by the generals for its mobility and rapidity in opening fire and was employed on all possible occasions. It was particularly useful against the Austrian cavalry, the scanty French squadrons being hopelessly inadequate. Daring, boisterous and quarrelsome, with a daredevil aggressiveness and an insufferably superior attitude, this corps provided the ideal career for such colourful characters as Debelle, Foy, the irascible Mossel, Sorbier with his reputation of bravery and Seruzier, a braggart behind his big moustache.

Apart from the artillery attached to the battalions of Volunteers, deemed ineffectual and even detrimental because of wastage in men and material, but

Ypres, June 1794. At the beginning of the month, a French column bombarded the town. A few days later, General Pichegru opened the siege. The town fell on 16 June, and the French took 6,000 prisoners. In this painting by Felix Philippoteaux, the French gunners are celebrating the announcement by their General of the fall of the town.

Roveredo-Calliano, 4 September 1796. The guns of General Dommartin have shaken the resistance of the Imperial army entrenched in the fortified village of Calliano.

Arcole, 15-17 September 1796. During the battle the gunners, in the heat of action, fired at point-blank range. The Austrians are in the foreground, the French are crossing the bridge.

kept in being for political reasons, the field artillery's weak point was the organization of its transport, entrusted to civilian companies such as Baudouin and Lanchère. The use of horse artillery and the adoption of daring tactics obliged these civilian carters to venture out under enemy fire. Riots were commonplace, equipment frequently abandoned and lost because "Lanchère's hussars", ill-treated, undernourished and the butt of jeering and mockery, were reluctant heroes. In 1794, Major Eblé revolted against this system, in force since the Middle Ages.

The law passed on 7 May 1795 concerning artillery organization did not improve matters in this respect. The Military School at Châlons, where famous men such as Duroc, Drouot, Marmont and Ruty had been trained, was maintained but a new *Ecole Polytechnique* was opened for future artillery officers. The number of horses regiments was reduced to 8, a new battalion of pontoniers was formed and the total strength of the arm brought up to 20,000 men. At the end of the year, the artillery-park comprised 4,816 bronze guns for siege warfare and the garrisons, 2,851 iron cannon for garrisons and coastal defence and 2,543 field guns. But it lacked 30,000 horses and this limited its use in the field.

Development in artillery tactics remained confused. Little attention was paid to the precepts of Guibert and du Teil regarding the large-scale use of batteries. This may have been because they were so unevenly distributed over the divisions, and because of the wide variety of calibres. Moreover, battalion artillery was not too well suited for use as a weapon for massive assault. Finally, the generals of the old regime clung stubbornly to their old customs while those who had risen to the same rank through the Revolution knew as little about Guibert as about the overall tactical aspects of artillery. At Kaiserslautern, Hoche was at a loss to know how to deploy his cannon, still vastly preferring the bayonet attack. Kléber, on the other hand, used his with masterly skill at Fleurus and, at Kostheim, outside Mainz, he concentrated the fire from his three divisional batteries on to the flank of the Austrian columns and forced the enemy to retreat towards its fortified positions.

Although the French were slow to adopt the tactics of concentrated offensives, the Austrians were even less enterprising in this respect. At Jemappes, Neerwinden, Wattignies, they retained their traditional defensive posture. But the offensive spirit advocated by du Teil was enthusiastically taken up by the horse artillery. Sometimes, this was carried to an excessive degree as in the case of a reckless battery attached to the Balland division, annihilated at Wattignies before having even unlimbered its cannon. At Arlon, however, Captain Sorbier conducted the first ever artillery charge against a hitherto impenetrable Austrian square. It broke through, opening the way for the carabineers.

This form of combat, which depended more on temperament than discretion, was soon adopted by the Russian horse artillery which, in existence for some time, had recently been reorganized by Count Zoubov. Boasting military transport and an excellent remount system, this artillery consisted of batteries of seven 6-pounder cannon and seven large howitzers known as "unicorns". The Prussian horse artillery was rapidly enlarged from six to twenty batteries while, in Austria, the Hanoverian Captain Sharnhorst, whose mounted battery had distinguished itself at Hondschoote, insisted on units being increased from six to nine guns. The first regulation concerning Spanish horse artillery appeared in 1796. In 1794, the English raised a "Drivers Corps" to replace civilian drivers and hired horses. Until 1806 it served the entire army, but in that year it was renamed the Corps of Royal Artillery Drivers. It was not absorbed into the Royal Artillery until Waterloo in 1815. The first permanent force of artillery in England had been raised in 1716, and had had the title Royal Regiment of Artillery conferred in 1722. The regiment was then commanded by Colonel Albert Borgard, a Dane by birth.

Assuming command of the Italian Army in the spring of 1796, General Bonaparte brought to his new role the experience gained as an officer commanding the artillery in this army, his own intensive studies into military strategy and history, all he had learned from du Teil as well as from his own efforts in capturing Toulon.

100

On the Rhine, the winter of 1796-1797 was marked by the battles waged for possession of the bridge-heads over the river. At Neuwied in April 1797, Debelle and Sorbier used cross-fire on the Austrian redoubts and succeeded in destroying them. Shortly afterwards, the Directory abolished regimental artillery and replaced the civilian artillery waggon drivers by foot-soldiers.

In the meantime, Bonaparte was covering himself in glory. His predecessor, General Schérer, an indifferent strategist but an able tactician, had divided his artillery into "static cannon" for battering the enemy positions and "combat cannon" for supporting attacks. General Lespinasse, a former artillery commander in the Army of the Western Pyrenees, with the same appointment in the Italian Army, quickly gained the confidence of his young chief. Lespinasse had a large but anomalous assortment of siege and field pieces at his disposal and these were supplemented by Austrian 6-pounder and 12-pounder cannon procured in Turin. These corresponded in size to the French 5-pounders and 11-pounders, thus scarcely assisting the standardization of calibres that Bonaparte aimed to achieve.

Finally, after repeated requests, twelve pieces of horse artillery, eight 8-pounder cannon and four 6-inch howitzers, arrived under the orders of General Dommartin. "Major-generals should not oppose the dispositions of the General commanding the artillery", stated the Commander-in-Chief when issuing this order, thus conferring considerable authority on the latter while still continuing, himself, to divide his batteries up among the various divisions. Masséna and Augereau each had one heavy battery and one horse battery and this allocation was soon made to each of the army's six divisions and, apart from rare exceptions, remained the rule until the end of the Empire period. A reserve of thirty-six guns was placed at the disposal of the Commander-in-Chief. Each gun was supplied with 300 rounds, half of them being held in the supply depot. A section equipped with one gun and one howitzer formed part of Bonaparte's guard, which later developed into the Artillery of the Imperial Guard. Nothing is known about the role played by artil-

lery at Arcole and Rivoli but, at Lodi, the fire from thirty cannon set up on each side of the bridge over the 200-yard wide river Adda silenced the fourteen Austrian guns in action on the opposite bank and enabled the infantry to cross the river. On 5 August 1796, at Castiglione, Bonaparte launched an attack with 30,000 men against the 25,000 Austrians under General Wurmser who were drawn up between Solferino, on the right, and Monte Medolano, which bristled with cannon, on the left. Augereau and Masséna launched a frontal attack while Lespinasse's twelve guns remained on the right. In the meantime, Marmont's nineteen guns intervened brilliantly, silencing the artillery on Monte Medolano and supporting an attack on the flank and rear conducted by Verdier, Beaumont and Fiorella.

On 3 January 1800, a few weeks after Bonaparte had become First Consul, the "artillery train" was established, its duties being to harness and drive the vehicles of the corps. This was a tremendous improvement, although proving a long and complicated operation to put into practice.

On the Rhine, Moreau organized his army's artillery in divisions of two batteries while Bonaparte's Reserve Army was crossing the Great St Bernard Pass with ten guns attached to each of the four large units, as well as a reserve of three mounted companies including one from the Guard. Guns, guncarriages, wheels and ammunition boxes had to be dismantled and packed on to sledges made from tree trunks. Helpful at the crest and in passing the snow line, the heavily laden sledges had to be held in check when descending the other side of the pass. Running the gauntlet under the guns of Fort Bard was an epic feat. Nevertheless, the equipment did not arrive at full strength in the plain of Marengo where the Austrian army under General Melas attacked with one hundred pieces of artillery. By nightfall on 14 June, the French were in retreat under concentrated fire from the enemy artillery commanded by Colonel Reisner.

Summoned by Bonaparte, General Dessaix arrived at San Giuliano. "... Cannon, cannon! We must have cannon!" he confided to Marmont. He had brought up eight with the division under Boudet.

Camp of Boulogne, 1804. A battery of heavy mortars points out to sea, for defence against the daring raids of the blockading British fleets. The range of these mortars would not exceed about 1,700 yds.

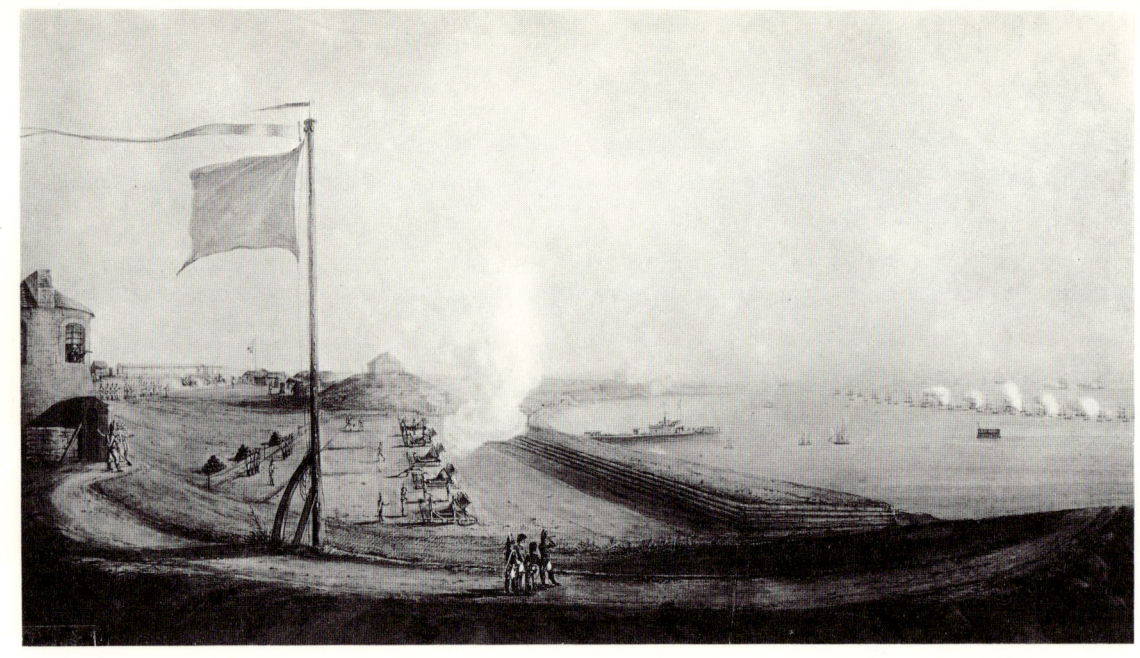

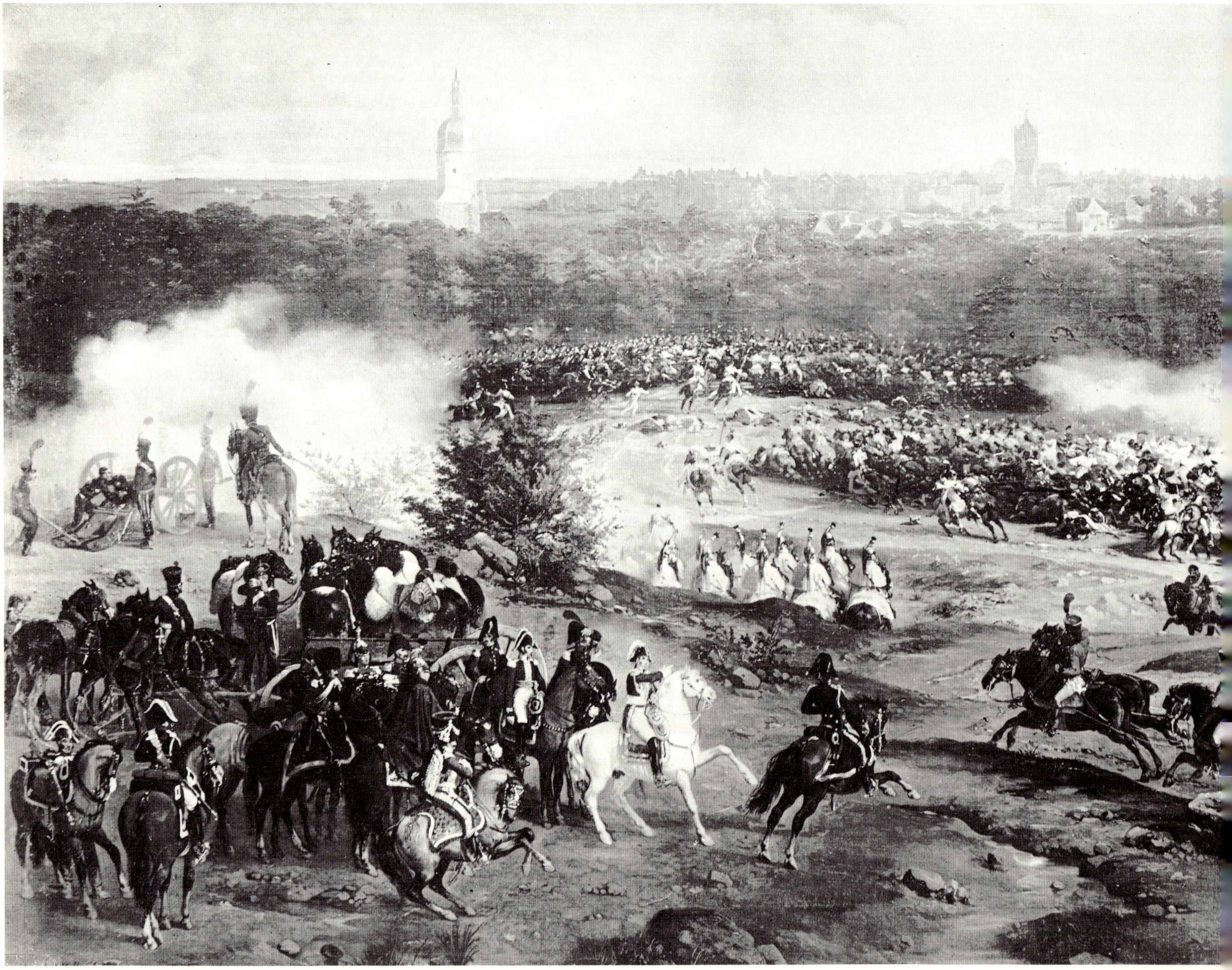

Landsberg, 11 October 1805. Napoleon imposed on his enemies a war of movement that affected the opposing artillery as well as the other arms. At Landsberg, Prince Ferdinand's artillery was surprised, caught, and overthrown by a charge of the 26th Chasseurs à cheval, and was not able to take part in the battle.

The Officer in Command of the artillery salvaged a further ten guns and, deploying the concentrated fire-power from his battery of eighteen guns, opened up a hellish gun-fire which caught the jubilant Austrian column at an angle as it approached from Marengo. Attacked by Boudet and Kellermann, the Austrians fled as far as Alessandria, pursued by the French cavalry and the searching cannon fire.

During the autumn campaign, Major-General Marmont, commanding the artillery of the Italian army, divided his 160 field guns into divisional artillery, reserve corps and army reserve. General Moreau concentrated his artillery at Hohenlinden while, at Moeskirch, a compact group of eighteen cannon was decimated by cross-fire from twenty-five guns which the Archduke Johan had divided into two batteries. The Austrian artillery, superior in numbers, also had excellent tactical skills.

Aicha-Augsburg, 9 October 1805. This picture by Jollivet shows a gun of the horse artillery advancing at a gallop to the support of Sebastiani's dragoons and the Hussars of Margaron. Excellently trained, the gunners and drivers of light artillery were noted for their dash under fire.

In France, increasing criticism was being levelled at the ordnance laid down by Gribeauval, although it had recently been adopted by Bavaria. The recent campaigns had shown up its shortcomings. *La commission extraordinaire du matériel d'artillerie*, under the chairmanship of Marmont, was commissioned by the First Consul, Bonaparte, to recommend any desirable alterations. First of all, this committee declared that "field guns should be as light and as mobile as is compatible with good service". In fact, making allowances for the wars and the limited metallurgical skills of his time, this is precisely what Gribeauval himself had sought to achieve. As the industry had made no progress since then, the committee confined itself to laying down general directives, leaving it to "experimental committees" to carry their ideas into practice. First of all, calibres were to be revised: the 4-pounder cannon, which was considered too light, and the 8-pounder, considered too heavy, were to be replaced by the 6-pounder already in use by the Austrian and Russian armies among others. France already had a considerable stock of these pieces, together with their gun-carriages and ammunition, having acquired them from the enemy in the course of various engagements.

At last, one major step had been taken towards the standardization of types. The only pieces of ordnance henceforth to be used in the field were the shortened 12-pounder, the 6-pounder and the 5½ inch howitzer. The 12-inch calibre cannon was retained in siege trains, together with the shortened 24-pounder and the 24-pounder mortar. Mountain equipment was to be formed. The number of types of wheels, axles, etc. was reduced. Studies and surveys were in progress but the experimental committees, pressed by the First Consul and the Minister, started production of the "short 6-pounder field gun, System An XI" in 1803 before detailed plans had been drawn up. This piece of ordnance had a calibre of 97.7 mm, a bronze barrel made from an alloy of copper and 10% tin, a length of approximately 6 ft and no ornamentation. The gun-carriage and limber were similar to those prescribed by Gribeauval. Bad planning, faults and miscalculations soon obliged the project to be scrapped and started afresh. However, the importance of artillery was increasing. On 3 April 1804, seventeen colonial companies were created for the defence of the colonies of Santo Domingo, Senegal, the West Indies and the Mascarene Islands.

At the time of the Empire, the French artillery comprised the artillery of the Imperial Guard, eight regiments of heavy artillery and six of horse artillery, eight transport battalions, two battalions of pontoniers, fifteen companies of artificers, thirteen companies of veterans, 130 coastal companies and four hundred officials for the maintenance and control of equipment. Now an independent arm, although working in conjunction with the other branches of the army, artillery had developed tactical skills and gained in speed and flexibility, mainly by reason of its offensive spirit, even if its equipment remained virtually the same.

In fact, the "strategic mobility", preached by the Baron Du Teil, Napoleon's tutor and officer commanding the Auxonne Military College, had now been achieved owing to improvements in the European roads, a reduction in the weight of equipment and a more judicious organization of the army into administrative staff, foundries, arsenals and artillery depots. Tactical mobility had also been achieved by means of unceasing technical improvements such as the introduction of limbers carrying a ready supply of ammunition, the militarization of drivers and horse-teams as well as innovations in methods of gun-laying, such as the elevating screw and the graduated, all-round tangent-scale or sight which considerably improved accuracy. The use of the cartridge, consisting of a cartridge-bag and the projectile fitted with its sabot, of the quick-match and portfire also made for greater regularity and reliability of fire.

However, the speed and precision of troop movements were mainly due to the rigorous training and discipline of officers and gunners who regularly practised, in dumb show, the numerous motions and tactical drill exercises with vigour and promptness. Between each exercise, the troops had to come to attention at their posts. There was only one word of command: "Charge!" Such automatism was only made possible by the excellent relations existing between officers and men. "A non-commissioned artillery officer should only become an officer through a brilliant feat of arms, or after eight years' service", wrote the Emperor to the Minister of War, thereby presupposing a period of ten to twelve years' service. "To have officers with only eight years' service since entering as soldiers is a very pernicious thing." Among this *corps d'élite*, of which the Emperor was later to say "great battles are won with artillery", the Footguards and Horseguards particularly distinguished themselves. Highly trained, they had extremely skilful gunlayers or Gun Captains and gunners who had fifteen or even twenty years' service. Champions of the competitions organized every year at La Fère between the best gunlayers from all regiments, they could fire three rounds per minute.

Tactical exercises were not determined by any particular rules or regulations but merely derived from long-established customs and practices. There was no shortage of technical works, memorandums and artillery manuals on the subject, not to mention notes from the Emperor on a wide range of topics. "Busy yourself to the utmost with the artillery", he wrote to Berthier, "we are still lagging behind in this respect and one can never have enough". He frequently stressed the need for having a great many cannon, "... even cannon for the infantry. Every day, I become more convinced of the harm done to our armies by abolishing the regimental guns..."

Writing to Prince Eugène, he defined his ideas on tactics. "... The cannon, like all other arms, should be combined in great number, if one wants to achieve important results..." At Austerlitz, with 18 heavy Austrian cannon and 139 pieces of Gribeauval ordnance divided among the infantry and cavalry divisions or held in reserve by the Imperial Guard, the Emperor massed 80 guns on the Pratzen plateau at nightfall and succeeded in routing the allied armies of Austria and Russia.

On the eve of the battle of Jena, by torch-light, Napoleon widened the road leading to Landgrafenberg and supervised the mounting of cannon and ammunition boxes, drawn by twelve horses, to the summit. The batteries were in position at sunrise on 14 October. That day, preceded by their cannon, the divisions commanded by Gazan and Suchet seized twenty Prussian guns and, with the ground cleared, the 10th Light Infantry went on to capture ten cannon attached to Hohenlohe's horse batteries.

At Auerstaedt on the very same day Davout, with only 40 pieces of divisional artillery in the morning, counted no fewer than 115 the same evening, owing to his appropriation of Prussian weapons.

In fact, the Prussian artillery, like its army, was in a sorry state. Brunswick, Tempelhof and Scharnhorst had appealed in vain to the royal officials and bureaucrats. Although its divisions were equipped with three batteries, including one horse battery, no assault artillery units were available to the Army.

Horse artillery was very much in vogue throughout Europe. Archduke Karl converted his mounted batteries into individual horse artillery companies and the German states of Saxony, Baden and Hanover followed suit. England had ten horse batteries but only brought them into use after the strictest, most intensive training. In Russia, horse artillery had been employed with great efficiency for some time. Cannon from the battalions were gradually detached from the infantry and transferred to the regiments of artillery. On 23 August 1806, the Russian artillery was commissioned as an independent force, the new administrative and technical units being formed into "brigades" of four to six batteries, one brigade attached to each division.

At Eylau, the Russian General Benningsen had 400 guns at his disposal, more than twice as many as Napoleon's forces. Taken by surprise, the French deployed twenty-four pieces of horse artillery attached to the Imperial Guard. Reinforced by the ordnance of the cavalry reserve and the 7th Corps, they checked counter-attacks launched by the enemy after Augereau's divisions were thrown back.

Although the use of horse artillery spread rapidly in Europe, concentrated assault tactics and the large-scale engagement of artillery were slower to gain ground.

Archduke Karl reported effective grouping of Austrian batteries at Rastadt and Friedberg. An explicit instruction was signed by him, issuing

Friedland, 14 June 1807, 6 pm. The battle started at 3 am. In the picture below, the French troops are in the background; on the left, the guns of Sénarmont have fired 3,500 shot into Friedland (in the foreground) and against the Russian batteries in the bend of the river. By the evening, Napoleon had beaten 69,000 Russians and taken 80 cannon and 54,000 soldiers.

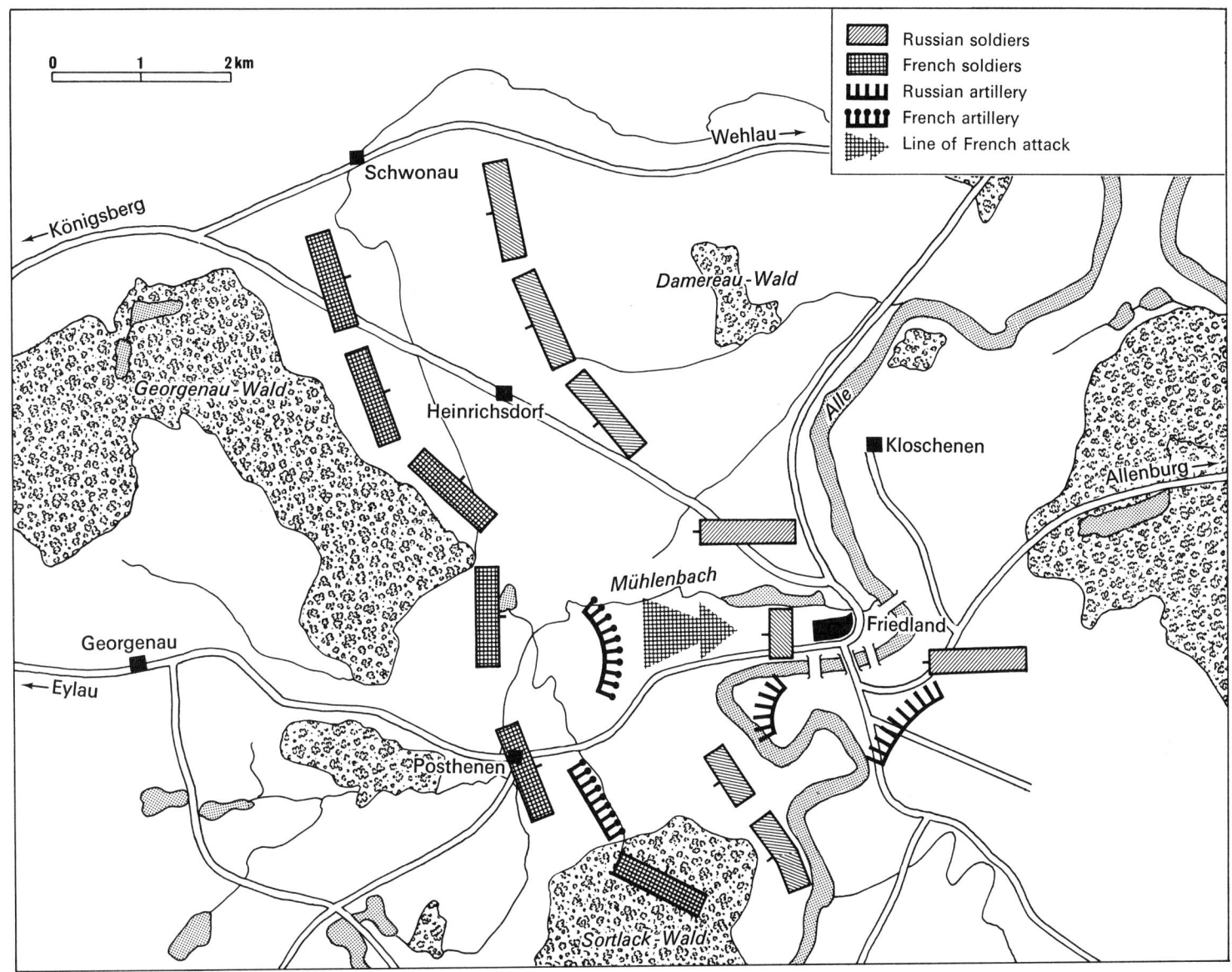

At Friedland, Sénarmont set up a grand battery of 30 guns, its right flank sesting on the Sortlack woods, and its left on the Mühlenbach. The Post-henen-Friedland road divided the guns into equal groups, whose fire was united to play on the Russian batteries on the right bank of the Alle, and to support the main attack on Friedland.

directives regarding "the massive use of artillery". Strangely enough, his own forces were victims of such deployment at Aspern and Essling. A work published in 1800 by General Lespinasse, *Essay on the Organization of Artillery*, was translated into German the following year and exerted considerable influence throughout European military circles.

At about this time, Songis was "Principal Inspector of the Artillery" in the French army. He had succeeded d'Aboville, a veteran of the American War and artillery commander at the Battle of Valmy, who had been appointed to this office after the death of Gribeauval. Songis was succeeded by Marmont whose ambition was too great to be content with the high command of the artillery. He relinquished the post which was subsequently occupied by Lariboisière, Eblé, Sorbier and a succession of others, all but one being officers sprung from the nobility of the old regime.

The traditions of the French artillery were still alive, and the innovations planned by Guibert and his followers were well on the way to fruition.

General Hureau de Sénarmont, commanding the artillery of the French 1st Corps, was a vigorous opponent of splitting up artillery, a method practised by the "irresolute and hesitant". He was an advocate of the principles laid down by the military experts of the eighteenth century, according to which artillery should act as one body both in defence and attack. He believed that batteries did not necessarily have to be drawn up in the same position as such a mass formation would, in fact, constitute a target too tempting for enemy gunners. In his opinion, they could be formed up into several groups, yet still be capable of concentrating their fire on the same objective. An instruction to this effect, worded by the General himself, was put into practice by the units under his command. Prudent, courageous and an excellent horseman, Sénarmont inspired in his gunners absolute confidence.

On 10 June, at Heilsberg, leading a strong advance guard, Murat launched a vigorous attack on Benningsen's eight divisions before even assembling his forces together. He suffered heavy losses and the cannonade lasted until midnight. Ill and indecisive, Benningsen, himself, escaped along the right bank of the river Alle and reached Friedland on 13 June.

On his arrival, the Emperor took Murat severely to task for his "erroneous manoeuvre" which had caused such "butchery". He marched to Heilsberg and ordered pursuit. Wishing to cut off the road to Koenigsberg, he dispatched Murat, Ney, Davout and the Imperial Guard in this direction. Soult, together with the 1st Corps under the command of Victor, proceeded by Landsberg to drive back the Prussians. Only the dragoons led by Latour-Maubourg and the division under General Lasalle followed the enemy along the right bank of the river.

On the evening of 12 June, the main body of the army reached Eylau and its outskirts. Uncertain of the intentions of the enemy who seemed likely to escape, the Emperor held back, as he had before the battle of Jena, and dispatched two advance guards: Murat and Soult in the direction of Koenigsberg, Lannes towards Friedland. Davout commanded a support corps which marched on Jesau. For his part, Benningsen believed the French army to be on the march for Koenigsberg, covered on its right by a flanking movement on the Alle. His idea was to winkle them out of their positions and attack Napoleon by the flank.

On the evening of 13 June, the Emperor learned that the Russians were emerging from Friedland and beating back Lanne's cavalry. He immediately ordered the Field-Marshal to the town, to stop the Russians' exodus at all costs. He then formed up the entire army behind his advance guard.

Supported by Mortier and the first cavalry divisions to arrive on the scene, Lannes held out for 4 hours with 10,000 men, later increased to 20,000, against 70,000 Russians and ensured the safe arrival of the rest of the army through the Posthenen gorge. Meanwhile, Benningsen, whose left wing was drawn up on the edge of Sortlack wood and his right on the banks of the Alle, allowed the battle to drag on with the river at his back.

In the distance, from Georgenau, a cry went up: "Long live the Emperor!" He called to Lannes, "I am bringing you the army". With a telescope to his eye, he scanned the horizon. He saw the river Alle, fifty yds wide with its steep, precipitous banks, the right bank rising upwards. Numerous Russian batteries positioned in the river bends to the south of Friedland swept the left bank. A permanent wooden bridge and three pontoon-bridges had afforded Benningsen's army access to the plateau where the ripening rye, swaying in the wind, was split from east to west by the Mühlenbach ravine. Hamlets were scattered over the landscape: Sortlack and its wood up-stream from Friedland, Posthenen on the road to Eylau, Heinrichsdorf on the road to Koenigsberg. The infantry under the command of Oudinot and Mortier and Grouchy's troopers performed prodigious feats. Cleverly making use of the terrain, they would vanish on the right to reappear on the left. Daunted, Benningsen, Gortschakof and Bagration had launched violent but uncoordinated attacks. The battle had degenerated into an artillery duel. Lannes was congratulated.

By this time, the heavy cavalry had arrived, together with the cavalry of the Guard, the rest of the 8th Corps under Mortier, the lst and 6th Corps under Victor and Ney, as well as the footguards. The battle was about to take a different turn.

Benningsen's only line of withdrawal lay through the back of Friedland. It was therefore necessary for him to be ousted from the town by quickly shifting the position of the battle front. The right wing was to press forward, seize the bridges and hurl the enemy troops into the water.

The battle order took shape. On the right, Ney to the south of Posthenen with Latour-Maubourg at his command; Lannes and Nansouty between Posthenen and Heinrichsdorf wood; Mortier in this village and beyond with all the cavalry apart from La Houssaye's which was in reserve, with Victor and the Guard to the south of Posthenen.

At five o'clock in the morning, three volleys thundered out from twenty guns in action near this village. The 6th Corps broke out from Sortlack wood in two lines of ten battalions, 800 yds long and 75 yds wide, to be caught under oblique fire from the Russian batteries formed up in the river bend south of Friedland. The Emperor sent Dupont's division of the lst Corps in support but these were soon under attack by the Russian Guard.

Bagration's infantry advanced into the breach made by his artillery. Then Sénarmont intervened. With General Victor's agreement, he took command of the army corps' thirty-six guns, held six in reserve at Posthenen and placed the remainder on the other side of the hillock sheltering the lst Corps, the left wing at Mühlenbach and the right at Sortlack wood. These two groups, on either side of the road from Posthenen to Friedland, were to employ cross-fire to facilitate attack by the Army corps and to counter fire on to the Russian artillery on the right bank.

While the 6th Corps was in action, the right-hand battery of fifteen guns advanced to a position just out of range of the cannon at Friedland. The left battery, which also consisted of fifteen guns, rapidly took up position 400 yds away from the enemy. After firing five or six volleys, it advanced a further 200 yds and opened a running fire on the Russian batteries and the bridges. Bagration pulled his infantry back while Sénarmont pushed his guns to

only 120 yds and began to fire case-shot into the crowds of fleeing foot-soldiers and into the ranks of the cavalry which had no time to deploy. Ney immediately set off and drove the Russians out of Friedland. The Emperor gave the signal for the main attack and Mortier, together with the squadrons commanded by Grouchy, precipitated Gortschakof's troops into the river. The fighting had been fierce and relentless and the French artillery had fired no fewer than 3,500 rounds. Remarkably well deployed, it had been divided into tactical groups so as to operate as one body in front of the infantry for whom it had cleared the way. Sénarmont's daring, his thorough command of the arm and knowledge of its possibilities, as well as his skilful handling of the horse artillery, had clearly demonstrated the increasing value and importance of artillery. "... Sénarmont, you alone were responsible for my success", said the Emperor.

* * *

After the battle of Friedland and the peace treaty negotiated at Tilsit, Napoleon turned his attention to England and his plan to ruin British commerce with the abortive "Continental System". However, neither Spain nor Portugal complied with his embargo and it was decided to send troops to the Iberian Peninsula.

Through the negligence of its Government, the Spanish army numbered only 100,000 men. Its artillery consisted of four regiments with ten batteries of six guns, six of these being horse batteries. There were seventeen fortress companies and five companies of artificers. General Morla had prevailed on the Government to constitute horse batteries even before these existed in France. The Gribeauval method had been applied to the Spanish ordnance but lack of organization and insufficient resources resulted in it being too ineffective to be of much assistance to the other branches of the army.

Crossing Spain on their way to Portugal, the French troops caused great agitation. The Spanish King abdicated in favour of his son Ferdinand who, in turn, was forced to abdicate by Napoleon who gave the crown to his own brother, Joseph Bonaparte. On 14 July 1808, Bessières succeeded in opening the road from Madrid to Médina-del-rio-seco. Very soon afterwards, however, Dupont capitulated at Baylen to General Castaños on 22 July and, on 30 August, Junot surrendered at Cintra after having been defeated at Vimeiro by an Englishman, the future Duke of Wellington. Threatened both on the front and the flank, Joseph left Madrid and the French troops fell back to the river Ebro.

Napoleon himself arrived with 80,000 men, bringing the total strength of his army up to 200,000 men, supported by 300 guns. Lariboisière was appointed Commander-in-Chief of the Artillery. The artillery of the various army corps was put under the command of the same Sénarmont of Friedland fame, the brave and faithful Bourgeat, the former artillery commander of the Imperial Guard, Couin, Lauriston and Bicquilley, who had been seriously wounded at Friedland.

The Emperor faced 120,000 Spanish, Portuguese and English troops. On 10 November, Soult broke through the centre at Burgos. On the 11th, Victor breached the left wing at Espinosa and, on the 23rd, Lannes broke through the right at Tudela. The Emperor marched towards Madrid, forced his way through the mountain pass at Somo Sierra on 3 December, and brought Joseph back to his erstwhile capital. He then learned of the arrival of the English General, Sir John Moore, who was advancing from Portugal with 5 divisions, 30,000 men and 66 cannon. Leaving Salamanca, Sir John Moore marched towards the Ebro with the intention of cutting off the French retreat and threatening their communications. But Soult, dispatched by the Emperor, was hot on his heels. Sir John Moore conducted his famous retreat towards his ships at Corunna but he was killed during a last battle on 16 January 1809.

Napoleon heard the news six days later, near Burgos, on his way back to Paris, having been obliged to return with the news of the mobilization of the Austrian army, and rumours of cracks in the Empire. Intending to return, he left 200,000 excellent soldiers and 500 guns with his brother

Joseph, now restored to his very precarious throne. The Peninsular War became at once the training ground for the British Army, and a running sore in Napoleon's side. Wellington, by a series of carefully prepared strokes, kept the French forces fully engaged in an unsuccessful attempt to oust him from Spain and Portugal. After the famous Siege of Torres Vedras, Wellington seized the initiative, and inexorably pressed the French back across the borders of France. As battle succeeded battle, the superiority of the British line over the French column became manifest, and the firm hand which Wellington kept on his army formed a contrast with the actions of the Marshals, who, jealous of each other, did not support each other in any strong measure. Many of the devices learned in the Indian wars were employed by the British, amongst them mule-borne mountain artillery, which proved very suitable for the Spanish terrain. For an artilleryman, though, perhaps one of the proudest moments of the whole Peninsular campaign came at the battle of Fuentes del Onoro in 1811. Captain Norman Ramsay found himself and his Horse Artillery completely surrounded and cut off by the French cavalry. He and his men were presumed lost or taken by the rest of the army. Ramsay, however, charged the cavalry with his unit, and crashed through the surprised cavalry at a full gallop to rejoin the British forces without the loss of a single man, horse or gun. This episode is unique in the history of warfare.

Arriving at the Tuileries on 23 January, Napoleon found there was no time to lose. He upbraided the European diplomats in no uncertain terms and unleashed a storm of abuse at Metternich, the Austrian ambassador. Then he set feverishly to work reorganizing his forces. The Italian Army under Prince Eugène was strengthened and increased to 7 divisions of 85,000 men and 102 guns. A new Grand Army was formed and its artillery placed under the command of General Songis, as in 1805. Savoir was given the command of the Imperial Guard, Lannes the 2nd Corps, Davout the 3rd, and Masséna the 4th. The 7th Bavarian Corps was under the command of Lefevre, the 9th Saxon Corps under

Bernadotte and the 10th Westphalian Corps under Jérôme. A reserve of cavalry and artillery was established. Sailors from the Imperial Guard and Boulogne reinforced the pontoniers. In all, a total force of 200,000 men and 340 pieces of artillery had been mobilized and was at readiness.

This impressive muster was, however, paltry compared to that of the other European powers. Austria had 340,000 men and 200,000 reserves available, as well as 890 guns; her 11 army corps each with 3 divisions being split up into 3 armies. The German army under Archduke Karl had 200,000 men and 618 guns. In Poland, Archduke Ferdinand mobilized 37,000 men and 94 guns while the Italian army, under Chastener, had 50,000 men and 180 guns. The artillery was excellent, superior in numbers and fire-power to the French, although less mobile. The Archduke Karl's reforms were only just coming into effect, despite the anxiety of

the Emperor Franz. Archduke Karl started the war by crossing the River Inn on 9 April. In five days, between 17-23, Napoleon and Davout won five victories but most of the Austrian army had escaped in retreat towards Bohemia. The Grand Army marched on to Vienna, entering the city on 13 May.

The Danube had to be crossed in full view of Archduke Karl who had arrived in Vienna and occupied the Marchfeld. Napoleon decided to cross at a point opposite Lobau island but the river was rising at that time of year and the floating debris thrown into the fast current by the enemy troops carried away the bridges pushed out to the left bank, leaving Masséna's 4th Corps and Lannes' 2nd Corps stranded in Aspern and Essling. Lannes was mortally wounded and his troops withdrew with great difficulty to Lobau island under a terrifying barrage of artillery fire.

Battle of Wagram, 5 July 1809, about 10 am. Napoleon's first object was the rapid deployment of the army on the left bank of the Danube (right). The hundred guns massed on Lobau island pounded the left bank, giving cover to the assault of the infantry on left bank below Enzersdorf. The army deployed on the plateau of Raschdorf (middle left), where the Austrian army was lying in force in front of Aspern.

Battle of Wagram, 6 July, between 10.30 am (above) and 12.30 am (below). The great battery (centre right) advances by successive leaps, under the fire of the Archduke Karl's guns, Lauriston's gunners driving back the Hungarian grenadiers and cuirassiers. Behind this shield, Napoleon assembled an assault formation of 67,000 men. (bottom centre). The Austrians' right centre was beaten. After 23 hours of fighting the French fired 100,000 rounds, and decided the outcome of a difficult battle.

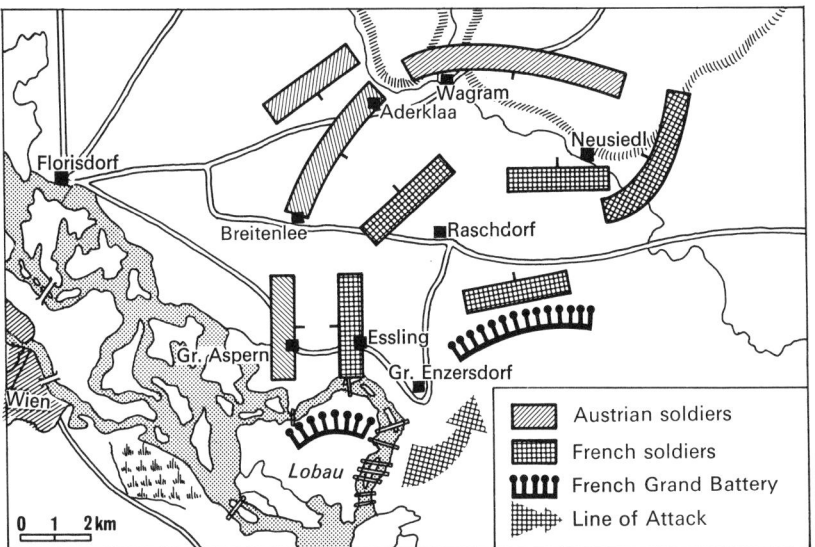

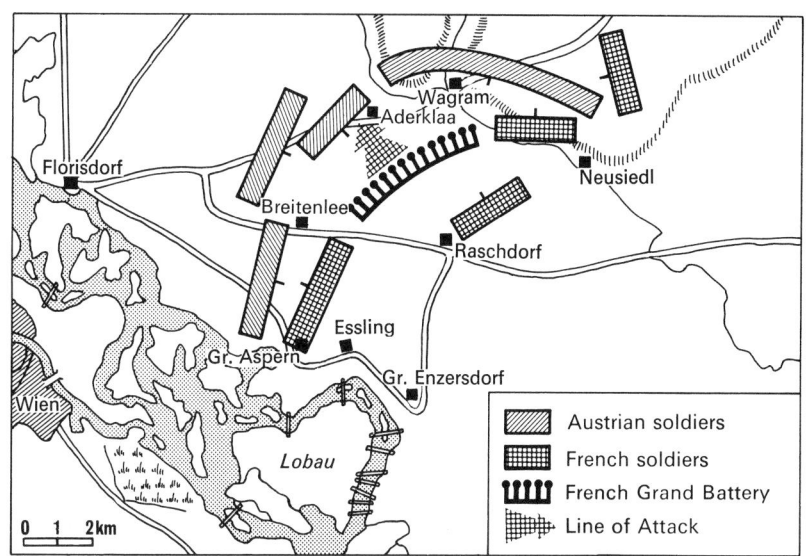

On the 4 June, the French occupied the Island of Lobau in force as their jumping-off point. In front of them, between Aspern and Probsdorf, the Austrians were strongly entrenched. The following morning, protected by the fire and concealed by the smoke of 109 pieces of artillery, the French crossed the Danube.

On 6 July, the battle restarted at dawn. Towards the end of the morning, to fill a gap in the centre of his lines and to gain time to form up his troops, Napoleon gathered a hundred guns. The gunners, commanded by Lauriston, drove back the Austrians. This was considered a marvel of engineering skill.

The battle of Wagram that followed showed Napoleon at his best, combining the talents of his three generals, Masséna, Davout and Oudinot, his superior and well-tried army, and his excellent artillery and engineering resources to form a massive and effective instrument of war. The Emperor was to feel the loss of Lannes, one of the most dependable and experienced of his Marshals, but the individual dash and initiative of various soldiers and junior officers in all branches of the service contributed to the victory.

On the Austrian side, the Archduke Karl's army was anxiously awaiting the Archduke Johann's army from Hungary, where it had been forced to retreat, but these reinforcements did not arrive in time for the battle.

Before and during the battle of Essling on 21-22 May, 52 guns from the 2nd Corps, 64 from the 4th Corps and 24 from the cavalry reserve, together with 101 chests of ammunition, had crossed the first stream of the divided river but only some of these had been transported over the second channel.

Impressed by the destruction wrought in his ranks by the Austrian artillery, superior to his own, the Emperor reverted to a system of regimental artillery, and distributed some of the three and four-pounder cannon which had been found in the arsenals at Vienna to his infantry troops.

Simultaneously, he reinforced his divisional artillery, mobilized the heavy guns and transformed Lobau island into an arsenal and industrial area with lighted streets, stores and workshops. He also speeded up preparations for the river crossing.

There were two main objectives: to link Lobau island to the right bank of the Danube, and to secure the army's rapid arrival on the left bank. Napoleon strode about every day with his principal officers, Berthier, Bertrand, Songis, Lariboisière and Major Dessales, in command of nine companies of pontoniers. They surveyed the work in progress, carried out inspections, made decisions, issued orders. A bridge on piles with a pier, pontoon-bridges, foot-bridges, large ferries and pile drivers were built. As the Austrians had entrenched between Aspern and Probsdorf, a bridge was thrown in this direction on 3 June with the idea of throwing the enemy still further off the scent. For, in fact, on 2 July, preparations were complete for the army to cross the river below Enzersdorf. On the evening of 4 July, the whole army assembled on the island, and the operation was to be carried out that very night. The rain was torrential.

At 9.30 am, Oudinot's Corps crossed Captain Larue's bridge while the brigade commanded by Conroux crossed the river in the ferries. The entire line was firing and Enzersdorf was burning.

At eleven o'clock the Emperor, gloomy and silent, waited with Berthier to be notified that the 180 yd long bridge, constructed in a single section, was ready to be thrown across from the island to the far bank.

"How long do you need to throw the bridge?" he asked Captain Heckmann of the 1st Pontoniers.

"A quarter of an hour, Sire".

"I give you five minutes! Berthier, your watch!"

Pistol in his right hand, Heckmann gave the signal to swing out, ready to blow out his brains if the operation failed. The bridge moved a few yards and came to rest under heavy rifle-fire. Heckmann had won the Golden Cross of the Legion of Honour. The divisions under Boudet and Carra Saint-Cyr were already on the left bank and the long-range guns were raking the plain.

By noon, 188,000 men and 488 guns were in position. The Danube had been conquered. But there was still the Archduke to contend with. His troops were not entrenched at Aspern, as had been thought, but were further back on the Gerasdorf-Neusiedel plateau. Napoleon's corps fanned out over the Rachsdorf plain while artillery salvoes paved the way for the divisions to advance behind the covering fire. Backed up by a fierce cannonade from Grouchy's horse batteries, Masséna captured Enzersdorf. Fires broke out among the Austrian positions, their batteries having taken severe punishment from the French 12-pounders. Covered by their artillery, regiments and divisions advanced amidst the most appalling uproar. However, the large units paid little heed to their formation and the direction in which they were heading so that the army was practically facing north by evening, and the Emperor could clearly discern the Archduke's positions on the Russbach plateau.

At about seven o'clock in the evening, he dispatched his leading corps against the heart of the enemy's positions at Wagram. Oudinot's 2nd Corps

with 148 guns was deployed on the left and Davout's 3rd Corps with 161 pieces on the right. In the centre, Macdonald with part of the Army of Italy suffered a crushing defeat. Blocked by both frontal and flank attacks, he was forced to withdraw over the Russbach plateau again, the Saxon troops and Lamarque's division being thrown into disarray. Finally, the infantry and the cannon of the Imperial Guard saved the day. The gunfire died away, but the artillery passed the night with its guns fixed to drag-ropes as the Austrians were on the alert.

At four o'clock on the morning of 6 July, cannon fire sounded the reveille. Both armies launched simultaneous attacks. The Archduke brought both wings into action while Rosenberg feigned an attack on his left against Davout. In the meantime, Hohenzollern and Klenau on the right endeavoured to cut Masséna off from the Danube. Boudet's division lost Essling, Aspern and their artillery, two pieces falling into the hands of Marulaz's cavalry. In trying to come to Boudet's assistance, the Duke of Rivoli was ousted from Adreklaa so that hordes of Austrian cavalry poured into the breach, enlarged still further by panicking Saxons under Bernadotte.

"Ten thousand cannon balls over there!" said the Emperor to Drouot.

Six heavy batteries and six horse batteries from the Guard, as well as five batteries from the line, arrived at full speed. Each battery in column was deployed on the order "Forward, in battery!", and opened fire. Over a front 2,000 yds wide, 100 guns commanded by Lauriston halted and drove back the Hungarian grenadiers and cuirassiers to clear a path for General Macdonald, leading the Army of Italy and followed by the Imperial Guard and the cavalry. The Austrian centre right was routed while Davout, supported by sixty guns formed up in groups, outflanked the Archduke on the left and Oudinot seized Wagram.

The decisive attack, backed up by artillery charges, brought about the Archduke's retreat, pursued half-heartedly by an exhausted army. Some 1,100 guns had thundered out for nearly 40 hours. The French had fired 100,000 rounds and Lariboisière, who paid five sous for every cannon ball brought back to the park, had retrieved 26,000. Tactics had evolved. Stronger storming columns were now necessary, as well as batteries operating in a single body. Napoleon realized that he no longer had the necessary superiority in artillery over possible enemies. The "Committee of Artillery Equipment" decided to adopt larger calibres and to build heavy 8-inch howitzers, similar to the Russian "unicorns." The first colossal howitzer was cast in Seville in 1811 and used at the siege of Cadiz.

As Emperor and a conquering hero in the Roman mould, Napoleon wanted to settle the question of the Bosphorus and Constantinople, and to force the Russians back on to their steppes.

His wish was prompted by the Tsar Alexander who protested at the annexation of the North Sea coasts by France, the occupation of the Grand Duchy of Oldenburg and, particularly, at the Continental System. Fate was pitting the heir to Rome against the "Emperor of the Barbarians".

Apart from the numerous troops he maintained in Spain, Napoleon also disposed of forces capable of going into action in the theatre of his next operations. There was Marshal Davout with his reconnaissance corps on the Elbe; in Poland, there was the Polish army under Prince Poniatowski; there were auxiliary German contingents; on the North Sea coasts, there was the so-called "coast-guard" corps and, in Italy, Franco-Italian troops were maintained under Prince Eugène.

Under cover of these large units, the Emperor organized his forces. During 1810 and 1811, the number of infantry of the line and light infantry regiments were increased to 108 and 31 respectively. Two 3-pounders or 4-pounders and three ammunition chests were alloted to each regiment. These, together with a forge and two waggons, one serving as an ambulance, as well as two battalion coffers for bread and cartridges, were worked, harnessed and driven by the Company of Regimental Gunners. In 1811, the regiments of the lst Corps under Marshal Davout were issued with four guns and their ammunition waggons.

The Emperor deemed cannon necessary to reinforce his infantry, weaker than it had been in the past, but the effectiveness of cannon was still questioned. The guns were too dispersed to produce powerful effects while the companies of gunners were cumbersome and awkward to move, and guns remained idle in the second line or held in reserve.

In 1812, the mounted troops numbered eighty-eight regiments, each of five squadrons.

The Imperial artillery corps, augmented in 1810 and 1811 by Dutch formations, was made up of the following: nine regiments of heavy artillery, each with twenty-two companies; six regiments of horse artillery, each with seven companies with the exception of the 6th regiment which had eight; twenty-seven battalions of artillery train with six companies; seventeen companies of pontoniers; nineteen companies of artillery artificers and five companies of armourers. On a peace-time footing, these units represented 250 officers and 3,850 men and, on a war basis, 266 officers and 8,000 men. In addition, there were nineteen companies of "veteran gunners" comprising approximately 1,500 men, dispersed along the coasts, as well as 178 companies of 20,000 men acting as coastal gun crews.

The Corps also included forty-three "Artillery Boards", four general directorates for the manufacturing plant, forges, foundries and bridges, the latter two having been founded in 1811. It also included an instruction school at Metz, eleven regimental schools, depots, workshops, siege and garrison gunners. The Artillery of the Imperial Guard, under General Count Sorbier, comprised a general staff, four horse companies, ten companies of heavy guns including four made up from conscripts, a company of artificers, two battalions of artillery train and one company of veteran gunners. Apart from a few guns left in garrisons for training purposes, the entire equipment of the German armies was assembled in the arsenals at Wesel, Strasburg and Mainz.

As a rule, in the French army a company of artillery only joined forces with a company of artillery train, to serve six or eight guns, at such time as it took to the field. A company of heavy artillery or horse artillery on a war-time basis, together with a company of artillery train, formed an "artillery division", serving eight or ten guns

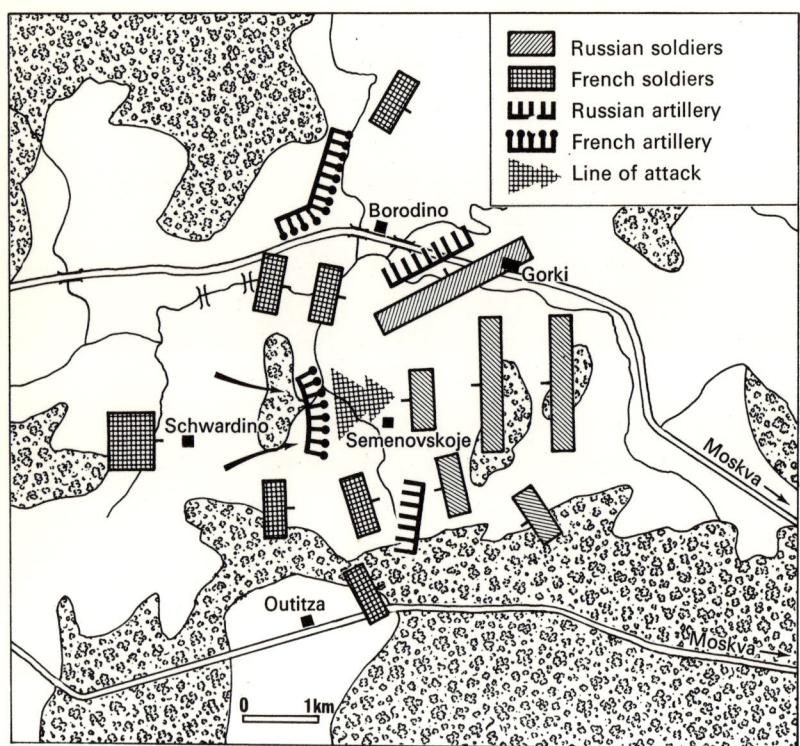

On 7 September, towards the end of the afternoon, Napoleon commanded Sorbier to regroup all available guns to assist in a last thrust in the very centre of the Russian position. Soon 200 cannon opened fire on the enemy columns; 200 new pieces were placed in battery. The Russians retreated under a hail of shot.

and accompanied by two or three ammunition waggons. Ammunition supplies consisted of 200 rounds per gun but, in the field, the Emperor doubled this supply to the entire artillery.

Each infantry division was alloted a company of heavy artillery comprising six 6-pounder cannon and two 5 ½ inch howitzers, as well as a horse company consisting of four 6-pounder cannon and two howitzers. Such a unit could proceed rapidly to a given point and open fire while awaiting the arrival of the division of heavy artillery which took longer to move. Two divisions of heavy artillery with six 12-pounder cannon and two 6 ½ inch howitzers, with an artillery park, formed the army corps reserve. One horse battery was allocated to each division of light cavalry and two to the divisions heavy cavalry. The artillery of the Imperial Guard, the army's supreme reserve, boasted 176 French guns comprising 23 regimental 4-pounders, 80 6-

pounders, 24 12-pounders and 40 howitzers, 8 of these being long-range pieces, as well as 32 pieces attached to the Italian Guard, making a grand total of 208 pieces of artillery. "In action", said the Emperor, "the artillery of the Guard provides for all contingencies". The main army depot held a supply of approximately half this number of guns in reserve. The General commanding the Army's artillery, Count Lariboisière, issued his technical and tactical orders to the Generals of the Army's artillery corps, a distinguished group of officers which included Pernety, commanding the lst Corps and a former officer serving the King; Dulauloy, who was succeeded by Aubry in command of the 2nd Corps; Fouché of the 3rd Corps; Breton, who had commanded the artillery at the siege of Saragossa and Anthouard de Vraincourt, commanding the 4th Corps, who had taken part in the siege of Toulon and the Egyptian campaign.

* * *

"I feel myself impelled towards an unknown goal", Napoleon once said. This dream evolved into his plan for conquering all Europe, an ambitious gamble in which he staked ten army corps, thirty-four divisions of French and foreign infantrymen and four cavalry corps, representing 500,000 bayonets, 80,000 sabres and 1,300 guns. Against him, the Tsar was to pit three Russian armies under Barclay, Bagration and Tormassof, nine army corps and six cavalry corps, representing 180,000 bayonets, 60,000 sabres and 15,000 Cossacks, as well as Europe's finest artillery. At the end of June, the 51-year old Barclay was at Vilna, guarding the road to Vitepsk with 130,000 men. Bagration, now 47 years old, was covering the route from Mohilef to Smolensk with 55,000 men. Tormassof was in Volhynie; a corps under the command of Wittgenstein was in training at Riga.

The Russian staff decided to remain on the defensive and refuse battle. Its armies would be withdrawn inside the country, the French drawn thither, exhausted, and then thrown out of Russian territory.

The Grand Army of the French was drawn up in three mass formations. On the right, Jérôme Napoleon, twenty-eight years old and lacking military experience, commanded the 5th, 7th and 8th Corps. In the centre, Prince Eugène, thirty-one years old and an indifferent strategist, commanded the 4th and 6th Corps. On the left, the lst, 2nd and 3rd Corps, the Imperial Guard, Murat and three cavalry corps were all under the direct command of the Emperor who, nevertheless, remained the Commander-in-Chief. On the extreme left were Macdonald and the Prussians and, on the extreme right, Schwarzenberg and the Austrians.

The material resources of headquarters and communications were not sufficient to coordinate the movements of several armies. It was nearly 190 miles from the Emperor's headquarters to those of Jérôme. With his astonishing military genius, Napoleon had obviously organized formations, built up a European army and worked out their strategic deployment. He thought of everything, made good every deficiency but there were limits even to his capacities. Nevertheless, he undoubtedly had the army's complete confidence.

On 24 June, the left wing of the army crossed the Niemen and this huge army was on the move. The Emperor entered Vilna on the 28th to find the Russians had withdrawn. Murat, Oudinot and Ney set off in pursuit of Barclay.

By now, the horses were eating nothing but green grass and were dying in their thousands. The artillery had been short of horses at the outset, as always at the beginning of a campaign, because Napoleon relied on obtaining fresh horses in the invaded country. This time, finding none, some of the cannon had to be left in Vilna. The Emperor searched for the enemy who had withdrawn behind the Dvina and, on 26 July, Murat only caught up with its rear-guard at Ostrowno. At Vitepsk, the enemy made as if to stop but then slipped away and reached

Borodino, 7 September 1812. Napoleon established 3 grand batteries facing the fortified Russian lines, using about 120 guns. In the early morning, the artillery opened fire, and the guns assisted the French in their assault on the great redoubt. Later the horse artillery, totalling a hundred pieces, advanced at a gallop to support Murat. At the end of the day, the Emperor combined all his available artillery, some 400 guns.

Smolensk to rejoin Bagration. Indeed, Prince Bagration too had eluded Davout and Jerôme. Attacked in the rear by Prince d'Eckmühl at Mohilev on 23 July, he reached Smolensk where the two Russian armies were reunited.

Two thousand miles from home, a third of the French army was sick or had deserted, looting wherever they went. Lacking both horses and drivers, guns and waggons littered the route. Napoleon tried to detain Barclay and Bagration on the road to Moscow, reached the river Dnieper at Orcha, urged Davout on towards Krasnoïe and marched to Smolensk on the left of the river with 175,000 men.

On 13 August, the artillery of the Guard fired salvoes merely to celebrate the forty-third anniversary of its formation. On 17 August, the old city was attacked with cannon but Bagration's soldiers disappeared amidst the flames and burnt the bridges behind them. For the fifth time in sixty days, the Russian Generals had refused battle.

However, the first objective had been achieved. Lithuania had been liberated and the old Poland conquered. The French should have stopped and waited for the arrival of spring and reinforcements. But on 17 August, Gouvion Saint-Cyr beat Wittgenstein at Polotsk and, shortly afterwards, Schwarzenberg defeated the army of Volhynie. The Emperor had set his heart on winning a victory as spectacular as that of Friedland under the walls of Moscow. And so the war went on.

On 5 September, Murat reported the enemy were on the Borodino plateau. Fortifications barred the road to Moscow. It appeared that, at last, the Russians were going to confront the French. Since public opinion would not allow Moscow to be surrendered to the French without a fight, the Tsar replaced Barclay with the seventy-year old, cunning and highly intelligent Kutusof. Finally, battle was to be given.

On that same day, supported by the batteries of the Guard, the 61st and 3rd line regiments stormed the Schwardino redoubt. Sub-lieutenant La Cène, attached to the 3rd line regiment, seized a Russian cannon, thereby partly compensating for the enemy's capture of two regimental guns.

For three days, 120,000 Russians had been entrenched in the hills running from north to south between Kolotscha and the forest of Outitza, barring the new road to Moscow. Everywhere there were breastworks and batteries, sometimes barely completed or simply in their first stage of construction. At the approaches to Borodino, a large redoubt with twenty-five heavy guns guarded the confluence of the Kolotscha and the steeply banked Semenofskoïe river. To the west of the village of Borodino, itself powerfully fortified and armed, were three redans shaped like arrow-heads, crammed with artillery. A few defensive works also lined the edges of the Outitza forest. Kutusof intended to check the French in this position, on a front almost two miles long and well covered on the right by the steeply sloping banks of the Kolotscha.

On 6 September, the Emperor surveyed the terrain. Davout suggested marching by night through the Outitza forest and falling on Kutusof's left flank the following morning, throwing his troops into the Moskova. Napoleon refused, unwilling to risk such a gamble. With 127,000 soldiers, even backed up by 580 guns, he could not carry out a large turning movement. He had barely enough forces to defeat the enemy by a frontal attack with the help of a pincer movement.

Barclay pressed his left flank on to the Great Redoubt and Bagration his right. Napoleon placed three large batteries, a total of 120 guns, behind the breastworks to demolish the Russian fortifications.

The night was cool and damp. The 6,500 soldiers of the Old Guard were drawn up in a square to guard the Emperor. In their bivouacs, the troops were eating horse-meat for the first time.

On 7 September at half-past five in the morning, a battery of the Guard gave the three classic knocks to ring up the curtain for a terrible cannonade. The three heavy batteries opened fire. One aimed at the three redans, the second the Great Redoubt and the third the village of Borodino. The right-hand battery, being too far from its objective, moved up into a closer position in the open.

Prince Eugène's Italian division stormed Borodino but was unable to force its way out the other side.

Davout attacked the three redans, defended by Voronzof's grenadiers. Despite murderous artillery fire, the Marshal heading the 57th succeeded in capturing the one on the right. The division commanded by Dessaix, withdrawn to the rear, was backed up by Marshal Ney who forced his way into the fortification, fighting like a grenadier. In the meantime, a division of the 3rd Corps seized the redan on the left.

A fierce Russian counter-attack on the two redans was checked by a formation of artillery made up of divisional batteries and those that had been held behind the breastworks. Then Murat intervened, leading Nansouty's cuirassiers. After several charges, six squares of the Russian guard were cut down with the sword and put to flight while the battalions under Ney and Davout seized the central redan under intensive fire from the Russian artillery. These battalions then proceeded forward and came to a stop on the outskirts of Semenofskoïe.

Leading Latour-Maubourg's 30 squadrons and sustained by a mass attack from 100 guns from all the cavalry's horse batteries charging at a gallop, Murat drove 2 Russian regiments back into Semenofskoïe and confronted 5 regiments of cuirassiers who counter-attacked.

At ten o'clock, artillery fire raged over a battle-front more than a mile wide while the division under Morand attacked the Great Redoubt, defended by the Paskievitch division. General Bonamy, heading the 30th regiment of the line, leapt inside the redoubt and held it with the help of the 17th regiment of the line and the 7th light infantry. The Russian front was broken. Ney and Murat sent a messenger to the Emperor at Schwardino, requesting reinforcements to turn the enemy retreat into a rout. Napoleon hesitated. Almost simultaneously, Kutusof was urged to send help to his generals who were dangerously threatened. The General Kutusof's artillery inundated Morand's division with cannon-balls and brought Paskievitch back into the battle, supported by the Likatchef division. The Great Redoubt was recaptured while, on the right wing of the French army, Ney and Davout were furiously attacked by Bagration who had sworn to recapture the redans or die in the attempt. They were saved by the arrival of the divisions under Marchand and Friant. In their regimental formations, the 15th light, the 33rd and 48th line regiments and Joseph Napoleon's Spanish regiment crossed the Semenofskoïe ravine. Deploying under hellish gunfire, they formed into squares of brigades to resist the repeated charges launched by the Russian cuirassiers. The fighting was terrific. "Bravo, Frenchmen", cried Bagration, who was killed a moment later.

"Soldiers of Friant, you are heroes!" exclaimed Murat who had taken momentary refuge in their midst. Davout was wounded. Threatened on their left by Poniatowski who had just captured Outitza, some Russian units began to fall slowly back.

The Emperor, who had been following the development of the engagement, now ordered Sorbier to combine the artillery of the army corps, as well as the artillery of the cavalry and the Guard, into a single battery. Summoning Junot, he ordered Murat's cavalry to take up its position on both sides of the Great Redoubt. Some 200 guns opened fire on the Russian columns. Eugène's infantry scaled the breastwork. Murat sounded the charge and Caulaincourt, succeeding Monbrun who had just been killed, together with Latour-Maubourg and Grouchy spurred on their squadrons. This produced confusion in the ranks of the horsemen of the Guard, resulting in a frightful mix-up. Under fire from 300 cannon, the Russian army fell back but their close columns steadfastly formed up to the rear of the battlefield seemed to defy the French. "Since they want more", Napoleon said to Sorbier, "give it to them". All the remaining artillery was put into action and 400 guns poured out a deadly fire. The Russians withdrew with the pride of wounded heroes. In all, 120,000 cannon shots had been fired and 80,000 men, including 30,000 Europeans, lay dead or wounded. "Never before", the Emperor said later, "did I see my army shine with so much merit".

The French army had indeed covered itself with glory. The infantry had been staunch and brilliantly commanded. The cavalry, performing heroically, had not been spared by the Emperor. The artillery was powerful, employed in massive batteries, operated by well-trained gunners and controlled by skilful and experienced Gun Captains. However, the Emperor had not displayed his customary ingenuity. His combat procedures, like those of the Russians, had been artlessly simple and straightforward. Battalions, cavalry squadrons and batteries had been launched into the fray, engaged on one side and the other until the final break-through and victory had been achieved.

It was victory, indeed, for the enemy was in retreat. A tactical victory won by exceptional troops, as well as a strategic victory since it opened the road to Moscow. A loudly acclaimed victory but, for all that, a Pyrrhic one.

* * *

As early as 1807, the Emperor had cursed "the plague of distances". He was well aware that the 1812 campaign beyond the River Niemen would create problems with the troops in the rear and, accordingly, had made careful preparations. Following the battle, he nevertheless remained behind at Vitepsk some time and only made a further bound forward after he had ascertained that his rear guard was properly organized. At Smolensk, he hazarded a long-distance march of 200 miles, certain of being able to engage the enemy in battle before Moscow. Flushed with victory, he massed his troops round the capital. Napoleon had 100,000 men in good condition, a complete and well-equipped artillery and the cavalry which was, however, now below full strength. But these forces were isolated, linked by a precarious route, fifteen days' march long, to the centre of operations in Smolensk which was still only partially prepared. The troops in the rear had not rejoined the advance guard.

With the onset of winter, the army was obliged to withdraw to its bases, but it was doubtful whether the rear was capable of receiving the advance troops for the winter season. Eleven days after setting out, intense cold paralysed the men while frost crippled the horses and held back the waggons. However, the Smolensk base was functioning, whatever the oft-quoted chronicles of Sergeant Bourgogne might have said on the subject. Comfortably billeted reinforcements joined up with the main army. The Emperor

ordered the material for two bridges which had been left at Orcha on the outward march to be burned. Some 600 horses joined the artillery of the Guard.

On 26 November, Oudinot's 2nd Corps and Victor's 9th Corps joined the army, doubling its total strength. On the same day, near the Berezina, the brigade under Badoise came across a convoy which had set out from Karlsruhe in July with provisions, shoes and equipment. Finally, about 350 guns with limbers crossed over General Eblé's bridges.

Then the drama began to draw to its tragic close, with 120 guns being abandoned at Vilna. The Polish corps managed to save a large part of its artillery. Although the Russians reported they had captured only 212 guns, Lariboisière and Eblé were lost. They died of cold and exhaustion in Koenigsberg on 21-31 December.

All Europe now rose against Napoleon. It was a duel to the death in a series of onslaughts by the "1813 conscripts", the dashing cavalrymen and a formidable artillery raised by General Evain, Chief of the Artillery Committee, consisting of four batteries to each division. In May 1813, 412 guns were limbered up and, by August, a further 1,200 were in action. Cannonade and victory at Lutzen. Cannonade, victory and pursuit at Bautzen. Cannonade from 800 guns and victory at Dresden. A French cannon ball killed General Moreau as he fought among the Russian ranks. But at Leipzig on 18 October, despite a cannonade from Marmont in the north, followed by an offensive launched by Drouot and the Guard with eighty guns, the Grand Army was in retreat towards Erfurt under fire from 2,000 guns. General Desvaux de Saint-Maurice, commanding the horse artillery of the Guard, rode at the head of his gunners. On the standard were proudly embroidered the names of all the battles: Vienna, Berlin, Madrid, Milan, Moscow, Warsaw, Venice and Cairo. Reaching the edge of Hanau forest, Drouot halted. Eighty guns were lined up in an attempt to throw Wrede's Bavarians into the Kintzig and delay the invasion for a few more days.

Alone against all Europe in arms, Napoleon held France at arm's length. The drama lasted sixty days. It was the epic of a handful of veterans and

youths, running here, there and everywhere in the icy winds of the Champagne region, using their speed and energy to surprise, defeat and madden two armies four or five times superior in number.

It was the feat of a Commander-in-Chief whose genius had never shone more brilliantly. At the battles of Champaubert, Montmirail and Vauchamps, the triumph of intelligence over force achieved victories that were as much psychological as military. Despite the fact that his troops were being inexorably thinned out, Napoleon's dread reputation intimidated his enemies so much that they actually withdrew at Craonne, Rheims and Arcis-sur-Aube. But the traitors' appeal to the enemy from Paris sabotaged these last desperate but brilliant manœuvres. The desertion of his marshals and the ingratitude of the French people brought about Napoleon's abdication and his departure to the Island of Elba in April 1814.

Escaping from the Mediterranean island and again established in Paris on 20 March 1815, Napoleon sought peace but the European sovereigns forced him

Battle of Waterloo, 18 June 1815. The allies obtained a decisive victory. In this picture by W.B. Wollen, the British cavalry are overrunning a French battery, stoutly defended by its gunners; 250 guns fell into the hands of the victors.

In the days following the battle of Waterloo, the inhabitants of the local villages became accustomed to the sight of passing French cannon captured in the battle. On the left of the picture is the house where Wellington made his HQ.

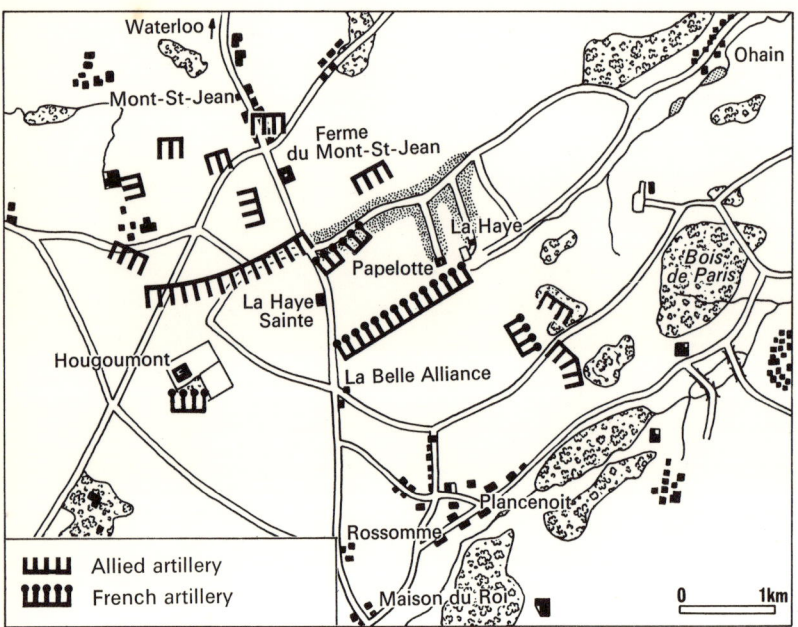

For reasons of clarity, this map does not give the artillery positions of the opposing armies. Wellington positioned his guns in a large open harrow formation, so that the French fire could not destroy them. The method of employing huge batteries as battering rams against an opponent had reached its limit.

back to war. He called France to the colours, mobilized the National Guards and the veterans, dispatched 400 naval guns, 6,000 marine gunners, 6,000 garrison gunners and 2,000 veterans to the empty and undefended garrisons to supplement the 11,000 artillerymen of the line serving the guns. He concentrated 300 guns in Paris, 100 of these forming mobile batteries. "We need cannon everywhere," he wrote, "we are fighting with cannon like one fights with one's fists!"

At the same time, the Emperor mustered 290,000 regular troops and 220,000 auxiliaries, posted an Observation Corps at each frontier and retained the "Army of the North" under his own command, facing Belgium. General Ruty commanded the Emperor's artillery.

On 14 June, the Army of the North, comprising 124,000 men and 374 guns, was grouped round Avesnes. It was commanded by the only four Marshals that remained loyal: Soult, the Chief-of-Staff; Davout, the War Minister; Grouchy, the cavalry commander, and Ney, who arrived on 16 June and was immediately given a command. Behind them were 160 Generals, excellent regimental officers and well-trained soldiers. Together, they formed a strong and vigorous army, fired with a passionate loyalty towards the Emperor. In its motley attire, this may have been a less splendid army than has been depicted by the poets and painters but, in grandeur and nobility, it transcended even the most romantic legends.

Concentrated around Beaumont on 15 June, the Army of the North confronted the Army of the Netherlands under the Duke of Wellington — somewhat weaker than it had been, for some of the veteran regiments that had achieved the successes of the Peninsular War had been shipped off to America — and the Army of the Lower Rhine under Marshal Blücher. Strung out on both sides of Charleroi, the Prussian troops extended as far as Liège and the English up to the North Sea. The Emperor intended to destroy them before the arrival of the other allied forces, which he hoped to defeat one after the other.

On 15 June, the French Army crossed the River Sambre at Charleroi on a front 3 miles long, an operation which was slowed up by various mishaps.

The Emperor missed the Prussian 1st Corps under Zieten, which withdrew to Fleurus-Sombreffe where Blücher constrained his three other corps to forced marches. Napoleon, apparently intending to proceed towards Brussels where the Duke of Wellington awaited him, sent Marshal Ney forward with the 2nd Corps under Reille and a cavalry brigade, followed by d'Erlon's 1st Corps and Kellermann's cuirassiers. These dispositions resulted in two battles being fought simultaneously on 16 June. The Emperor defeated the Prussians at Ligny with the rest of his army while, at Quatre-Bras, Ney fought a hard battle and only held Wellington's troops, which had hastened from Brussels, with difficulty. On 17 June, believing the Prussians to have been knocked out of the battle and in retreat towards Namur, the Emperor sent Grouchy in the direction of Gembloux with 32,000 men and 96 guns to pursue and finish them off while he, himself, intended to crush Wellington between his own forces and those of Ney at Quatre-Bras. At about noon, however, Wellington, fell back towards the north.

Setting out in pursuit with the 1st Corps and his cavalry, Napoleon arrived outside the inn called *A la Belle Alliance* in the evening. He surveyed his position. Before him lay the Hougoumont farm, the Bois de Paris, Mont-Saint-Jean and the village of Waterloo. Wellington's army had come to a halt in this village.

This somewhat miscellaneous army was composed of English and German mercenaries, Dutch militiamen and a few Belgians. However, its infantry was strong, its cavalry daring, and the artillery well-trained. Each infantry division was assigned a foot company or battery and a horse company or troop, each company having one or two howitzers and four or five 6-pounders or 9-pounders, the 9-pounders being of a new type.

A "troop" formed part of each cavalry brigade and a battery of four 18-pounders was held in the Army Reserve. As it so happened, a large corps had been detached to guard his lines of communication to Ostend, leaving the Duke only 67,000 men and 189 guns to do battle. However, from Wavre, where he had withdrawn his troops after the defeat at Ligny, Blücher, the Prussian commander, sent a message to Wellington that he would "attack Bonaparte's right flank tomorrow with two and perhaps three Army corps". This was a heroic effort. Before the defeat at Ligny on the 16th, his young army with

its excellent discipline had numbered 120,000 men, a battery of howitzers, twelve horse batteries of 6-pounders and twenty-five heavy batteries, seventeen consisting of 6-pounders and eight of 12-pounders. On 18 June, at dawn, the 4th Corps, commanded by Bülow and comprising 30,000 men and eighty guns, left its bivouacs near Wavre and started off towards Chapelle Saint-Lambert. The roads were so bad that the artillery had often to be carried.

Grouchy was completely unaware of the Prussians' move. His corps had arrived at Gembloux too late. In the meantime, under the protection of the 1st Corps, the Emperor's army approaching from Genappe took up its combat positions in the rain. It faced the Army of the Netherlands which was drawn up from the Saint-Jacques farm to Merbe-Braine, along the sunken road leading from Ohain to Braine l'Alleud. They were in a strong position typical of Wellington's major battles. The troops were protected by the reverse slopes of the high ground, yet could advance down to crush the opponents who mounted them: the Mont-Saint-Jean plateau, well covered on the flanks, inclined gently down to the forest of Soignes on one side and was steeply sloped on the side where the French were drawn up on ridges a 1,000 or 1,200 yds to the south. In the front line, Hougoumont, La Haye Sainte and Papelotte were strongly held. The English artillery was

Napoleon leaves his carriage after the battle of Waterloo. Tired, blind to events about him, he realised the bitterness of failure.

positioned ahead of the road on which the battalions were deployed, well screened by its banks and hedges and did an enormous amount of damage with its shrapnel shells, introduced in 1804, and usefully employing Congreve's rocket batteries, which although unpredictable, frightened enemy horses!

The reserve troops of the three armies were massed on the reverse slope, particularly behind the right-hand side where one flank was thrown back at right angles, Wellington expecting the French would attack on this side.

But this was not the Emperor's intention. He planned to attack La Haye-Sainte at nine o'clock with two divisions of the 1st Corps, backed up by twenty-four 12-pounders. Two other divisions, together with his reserves and the cavalry were to proceed towards Papelotte to drive the Army of the Netherlands towards the west, thus separating it from the Prussians. But his troops were slow to deploy and Drouot feared that the artillery would become bogged down in the sodden ground. The attack was delayed until eleven o'clock and Napoleon grew impatient, especially since a message received that night from Grouchy warned him that the Prussians were near Wavre.

At last, the French Army was in position. To the right of the Brussels road, on the Belle-Alliance ridge, were the 1st Corps and the Great Battery which had been made up to twenty-four guns with help from the Guard. To the left was the 2nd Corps. In the second line, Milhaud's 2,000 cuirassiers on the right and Kellermann's 3,000 on the left encircled the 6th Corps under Lobau and the cavalry under Domon and Subervie. Finally, horse guards were positioned on either side of the infantry and artillery at Rossonne. There was no longer any question of manoeuvring. The Emperor ordered Reille to display a show of force outside Hougoumont so as to induce Wellington to withdraw troops from his centre which Napoleon would then break up by concentrated fire from the Great Battery and a massive attack led by Ney.

As the first cannon fired, the English gunners looked at their watches. It was half-past eleven. At half-past one, a concentration of troops came into view on the right, near Chapelle Saint-Lambert. Bülow's Prussians were arriving. It was quite possible for the Emperor to have withdrawn or manoeuvred into a more favourable position. What would they have said in Paris? Napoleon had retreated? That was unthinkable! So vanity forced him to fight it out here, to stake his all in these twelve square miles. He ordered Lobau to hold the Prussians in check with the 6th Corps and the cavalry under Domon and Subervie. He then commanded the Great Battery to open fire on the English lines, and Ney to attack them.

On the road to Wavre, Grouchy heard the uproar. His generals Gérard and Exelmans wanted to make for the sound of guns but the Marshal stuck to his pursuit of the Prussians. By an irony of fate, Soult had just written to him saying: "Your move to Wavre is in accordance with the dispositions of S.A." But a postscript added: "An intercepted letter reveals that General Bülow plans to attack our right flank ... manoeuvre to join up with our right!" This letter, which was sent at about half-past two did not reach Grouchy until six or seven o'clock that evening. And, by then, it was too late.

To the right of the Brussels road, the four divisions of the 1st Corps advanced in close formation and reached the edge of the plateau. Here, they were thrown back by the English cavalry which was attacked by the flank and destroyed by the squadrons of the 1st Corps.

At three o'clock, the battle raged on indecisively around Hougoumont. In the centre, a second attack by the 1st Corps was repulsed. The Prussians remaining immobile, the Emperor hoped to have time to break through the English centre and gave Ney his cuirassiers. Twenty-four squadrons under Milhaud, backed up by twenty-five under Kellermann and the cavalry of the Guard, mounted the slopes in the face of the English artillery. Wellington, in anticipation of this, had developed an entirely new method of handling his artillery which was to prove of fundamental importance to the outcome of the battle. The infantry formed square to receive the French cavalry, with the guns deployed in open formation in front of the positions. The cavalry then ranged about the squares, achieving very little. Graham described it: "An intense fire of grape shot was maintained until the cavalry were right on the guns, when by the Duke's order, the gunners withdrew to the shelter of the squares: each time the cavalry recoiled from our unbroken infantry, the gunners again manned their guns and fired into the retiring enemy. The gunners of Captain Mercer's troop however never left their guns, and the cavalry never got through them..." The gunners would detach the wheels from their pieces to prevent the French carrying them off, and bowling them before them, would dive between the legs of the infantrymen. General Foy, one of Napoleon's finest artillerymen, said "The English gunners are distinguished from the other soldiers by their excellent spirit. In action their handling is skilful, their aim perfect, and their courage supreme".

Judging Wellington to be too busily engaged to be able to extricate himself, Blücher attacked Plancenoit, which was defended by Lobau and the Young Guard under Duhesme.

At Wavre, Grouchy's soldiers, who had come to grips with the 3rd Prussian Corps, had advanced slowly and laboriously and only reached the Rixensart plateau in the evening. As a result, Zieten had been able to proceed to Mont-Saint-Jean.

By seven o'clock, at Belle-Alliance, the Hougoumont farm was ablaze. Papelotte was in French hands and La Haye Sainte had been taken by Ney. A horse battery, brought up to the north of the farm at a gallop, opened fire 200 yds away from the English whose guns, limbers and gunners suffered. The Prince of Orange and Kempt, as well as many others, were wounded while Cumberland's hussars turned their horses' heads and galloped off, the colonel leading the way. "Hard pounding, this...", Wellington said to the men around him. Ney requested a few infantry troops to finish the enemy off. Napoleon had only nine battalions of the Guard left and time was running short because Zieten was approaching from Smohain and Captain du Barail of the 2nd Carabineers had just deserted to the enemy, informing Wellington that the Imperial Guard were about to attack. All the English cannon were immediately crammed with case-shot and

twice the charge. The English Guards, lying flat on their stomachs in the rye, waited with their fingers on the triggers.

At half-past seven, drums beating, the Guard was already on the march behind the Emperor. The band of the grenadiers played the *March of the Bearskins* by Gebauer.

In the first line, Friant, Roguet, Porret de Morvan, the 3rd and 4th grenadiers and four limbered guns commanded by Adjutant Duchand of the Artillery of the Guard rushed down the slopes of Belle-Alliance. On their left were General Michel and the 3rd and 4th Riflemen followed by three battalions of the Old Guard commanded by Generals Cambronne, Christiani and Martenot.

Behind them marched all that remained of Reille's battalions and the squadrons of cuirassiers. The phalanx reached the bottom of the ravine and started up the slope on the other side. The Great Battery which had been firing hard fell silent. Then the English cannon belched flames. Fifty guns spewed cannon-balls and case-shot fired by double charges. The Belgian General Chassé massed his units in preparation for the counter-attack. Emerging from the dead angle, the middle Guard, battered by the artillery, closed its ranks, pushing Brunswick's battalions out of position and advanced out of the range of the enemy guns. But Chassé's battalions threw the assailants back to the bottom of the slopes. On the left, the 3rd and 4th Riflemen were checked and decimated by the fire from the English Guards. Fifty officers and 1,200 men fell. Faithful to their tradition, Duchand's horse batteries had supported the infantry of the Guard up to a distance of only twenty-five yds from the English line. The attack had lasted twenty minutes. From La Haye Sainte to Papelotte, the jubilant cry went up: "The Guard is retreating!" The line of battle collapsed. Blücher's forces emerged from Plancenoit, singing a hymn of Luther's. The Guard was no more.

The victorious Wellington gave the signal for the advance. Some 250 guns were abandoned but Grouchy brought his own guns, as well as some captured from the enemy, together with his wounded and 28,000 fighting men back to France.

CONCLUSION

Thanks to the officers of the old regime, almost all the innovations planned by Gribeauval, Guibert, Feuquières, Bourcet and Dutheil had been carried into effect. The artillery was no longer a "professional body" but an independent military arm. Between 1792 and 1802, both the heaviest and the lightest guns were withdrawn from use in the field. Horse artillery, widely used on account of its mobility and the enterprising spirit of its commanders, became a special corps and in England, in February 1793, two troops of Royal Horse Artillery were formed, equipped with 6-pounders. As an after-effect, the infantry's heavy and cumbersome cannon were abolished. Between 1792 and 1796, the army divisions and corps received an allotment of divisional artillery while the artillery reserve, hitherto composed of several groups of men and material, became an "Artillery Reserve" in 1796 under Bonaparte, with mobile and powerful batteries under the control of the Commander-in-Chief.

The artillery commanders were "tactical leaders", whose offensive spirit showed a tendency to engage in close combat, to combine operational duties with other arms and to concentrate their fire on a chosen point. Developments in artillery were not concerned with technical aspects but with organization and tactics. Owing to the skill and energy of its commanders and the courage of the gunners, the French artillery victoriously asserted itself, despite the superiority in numbers of enemy batteries.

Nevertheless, judging the officers' training to be a particularly weak point, in 1801 the First Consul drew up notes on "A Draft Regulation for the School of Artillery and Engineering" which had been transferred to Metz. Once the officers had acquired the necessary theoretical instruction at the *Ecole Polytechnique*, the primary aim of their practical training was held to be "the knowledge and manoeuvre of all guns and infantry tactics".

During the few months that elapsed between the peace treaty signed at Amiens and the fresh outbreak of hostilities, the First Consul remodelled the artillery's equipment on similar lines to the Austrian type by choosing the 6-pounder cannon and the 5 ½-inch howitzer. But the continual succession of wars slowed up the adoption of this lighter weaponry.

The Emperor made few changes as far as the organization of artillery was concerned. Although he strengthened his infantry by bringing back regimental cannon in 1809, he never deviated from his oft-quoted preference for massive artillery attacks in battle, believing this to be the only means of achieving decisive results.

A reserve of mobile and powerful artillery was placed at the disposal of the army corps commanders and, first and foremost, of the Commander-in-Chief in order to go into action in a body under one and the same command. It was not necessary, or even desirable, for the cannon to be concentrated in dense groups. Although their fire should be concentrated on the same objective, the batteries themselves should be separate, operating from different directions so as to disperse the enemy's artillery fire. Artillery should be able to move about in the field, shifting from one firing position to another, as required, in order to support the other arms at the point where they were most needed.

The European powers at war with France, subjected to the superiority of her artillery, subsequently adopted the French methods of organization, the Gribeauval-type ordnance and its tactics. Although sheer weight of fire-power had enabled Russia, Prussia and Austria, with their efficient horse artilleries, to overpower the Emperor's gunners, they nevertheless suffered heavily from the effects of the Frenchmen's skill. At Grossbeeren in 1813, General Holzendorf had employed only sixty cannon, as had Lauriston at Wagram and the Emperor at Moskova. Until 1807, French artillery retained the lead but gradually lost its superiority as its adversaries adopted the principles advocated by Guibert and Napoleon himself. While incarcerated on the island of St Helena, he wrote: "Artillery, today, holds the key to the true destiny of armies and nations".

FROM WATERLOO
TO THE FRANCO-PRUSSIAN WAR
1815 - 1871

What gained the day for Napoleon at Friedland, Borodino, and elsewhere (though not at Waterloo!) was his tactic of deploying massive concentrations of field-guns, discharging cannister and grape shot at infantry ranges, rather than any superiority of the weapons themselves. French muskets, indeed, were of dubious quality, while the 6-pounder and 12-pounder cannon which wrought such havoc among the ranks of the enemy were the standard smooth-bore muzzle-loaders of the time. As such, these weapons exhibited many shortcomings, not least among which were a want of accuracy and a lack of propelling power—deficiencies which were common to guns of all three groups of the then prevailing Gribeauval classification which distinguished between land ordnance for coastal defence, for purposes of siege, and for use in the field. Of particular concern in weapons in this last category was the question of mobility, and from the time of Frederick the Great this had been answered, at any rate in part, by the emergence of horse artillery as an adjunct to the cavalry arm.

In the event, long overdue improvements to the guns themselves were not to be much in evidence before the middle of the century. Nor are the reasons for this far to seek. For one thing, the second (and final) enforced abdication of Bonaparte was accompanied by a determination on the part of the victorious allies that the settlements earlier reached at the Congress of Vienna should at all costs be maintained, so that for the next few decades the world experienced a period of comparative peace. For another, quite apart from the circumstance that, in the absence of serious strife among the main European contenders, there was little incentive for Governments to devote time and money to the development of new and more efficient arms; any such activity was further discouraged by a wave of economic depression which came in the wake of the recent conflict. And for a third, officialdom, both military and civil, tended to display an attitude of instinctive hostility towards new and unfamiliar ideas. As a prominent member of Great Britain's Ordnance Committee subsequently had occasion to explain, if only a single mistake was made in the course of every hundred decisions reached, this could be regarded as an outstandingly satisfactory achievement; and as ninety-nine out of every hundred inventions submitted for consideration were utterly worthless, even if he and his colleagues summarily dismissed every proposal they received, they could still claim to be doing remarkably well!

But more progressive influences were also at work. Industrialisation, motivated by a replacement of animal muscles by machines, had already reached the stage where these machines had been set to making yet other machines, a situation which promised to revolutionise production techniques, accompanied as it was by far-reaching discoveries in the realm of metallurgy, chemistry, and associated sciences. Among fire-arms, handguns were the first to feel the impact of these events, and so great was the improvement in their performance brought about by the progressive introduction of such features as percussion ignition, rifled barrels, elongated missiles, and breech-loading, that a like conversion of cannon, despite the difficulties such a programme would entail, could only be a matter of time.

The question of ignition provides a case in point. Although the British Navy pioneered the introduction of flintlocks for its ship broadsides as early as the middle of the previous century, use still continued to be made of the linstock (slow match) in conjunction with the portfire (quick match), an association in which the one, kept continuously burning, served to kindle the other, which was extinguished once the guns of a battery had been fired. This somewhat primitive procedure was not without its attendant dangers, and it is on record that in the heat of battle during the siege of Sebastopol, a gunner armed with a lighted portfire collided with a companion carrying an 80 lb bag of black powder, with disastrous consequences for all in the vicinity. But by this late date (1855), even the long established and much safer handgun equivalent, the flintlock (which generated sparks on demand), was itself slowly making way for an even more advanced device.

It so happened that several new explosive substances had recently been discovered, among them the fulminating silver of L.G. Brugnatelli. This laboratory curiosity was also linked with the name of France's leading chemist at the time of Napoleon, the famous C. L. Berthollet. It was he who was the first to isolate potassium chlorate, which he proposed to use in the preparation of a gunpowder substitute, claimed to be twice as powerful as the conventional mixture containing potassium nitrate. Yet a third detonating agent was produced by Charles Howard, when he experimented with the effect upon quicksilver of alcohol and aqua fortis, and so obtained mercury fulminate. Experience quickly showed, however, that all three substances were of far too sensitive and violent a nature to permit of their being used as the main charge in firearms (a mill at Essons making chlorate powder blew up, killing two people), and it remained for Alexander Forsyth, a Scottish clergyman with a fondness for duck shooting, to demonstrate that one or more of these compounds could be employed in a percussive capacity, *i.e.*, as a primer which would explode with a flash when struck.

After failing to awaken the interest of the British authorities, Forsyth undertook to develop his invention privately, and obtained a patent for his priming composition in 1807. From it was evolved the copper detonating cap attributed to Joshua Shaw, a device which soon found ever-increasing use in pistols and muskets. The application of percussion ignition to artillery, however, was a more gradual undertaking, in the course of which not a few expedients were tried, with varying success. Early attempts to substitute percussion for flintlocks on cannon encountered difficulties because of blast damage to the lanyard-operated hammer, a problem which Hiddens, an American, sought to overcome by making use of a hollow striker, the hole in which served as an escape vent. About the same time (1831) the Swedish Government introduced a system proposed by a Captain Collerstrom, in which the gun was fired by means of a blow aimed at a phial of sulphuric acid, whose contents inflamed the priming mixture on coming into contact with it. Other military authorities, though they retained both linstock and portfire as a standby, utilised a percussion tube and hammer device, later to be replaced by a friction tube, detonated by the withdrawal of a roughened bar.

The arrival of percussion and friction tubes was preceded by that of the revolutionary shell gun. In 1822, Henri J. Paixhans, a French artillery officer, expressed concern at the numerical superiority enjoyed by the British fleet, and advocated the use of horizontal shell fire as a means of redressing the balance. Trials conducted at Brest demonstrated the destructive capabilities of such missiles when aimed at wooden ships, and, after certain modifications had been carried out, the Paixhans gun was adopted by France and other maritime Continental powers. Great Britain reluctantly followed suit, not only in this, but also in the matter of the iron-clad warships which then appeared, in response to the clamour which arose in naval circles that some means be found to 'keep out the shells'. H.M.S. *Warrior*, the first British Iron-clad, was launched on the Thames in 1860.

Shell fire itself, of course, was no novelty. Such missiles, in the guise of hollow spheres of cast iron, filled with powder and fitted with a fuze, had been employed in land warfare for centuries. But their use had come to be restricted to the high angle fire

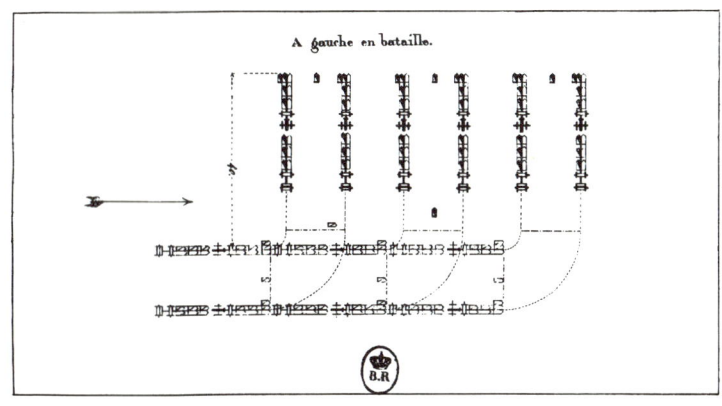

A gauche en bataille.

Gun-drill maintained the morale of the crew and fire-discipline. At the moment of firing everyone is in position. The man on the right swabbed out the gun and broke open the charge, the gunner at the left rammed home the charge and the ball, the gun captain, (right rear), assisted by the man behind with the hand spike, trained the gun, while the man on the left discharged it with his portfire.

Artillery became a service whose basic unit was the battery of about four guns or howitzers. The commander of a horse-battery kept up the flexibility and cohesion of his battery by exercising them in the normal evolutions in taking position under fire. Opposite, three plates from a French manual of 1836 on "the manoeuvres and evolutions of horse batteries".

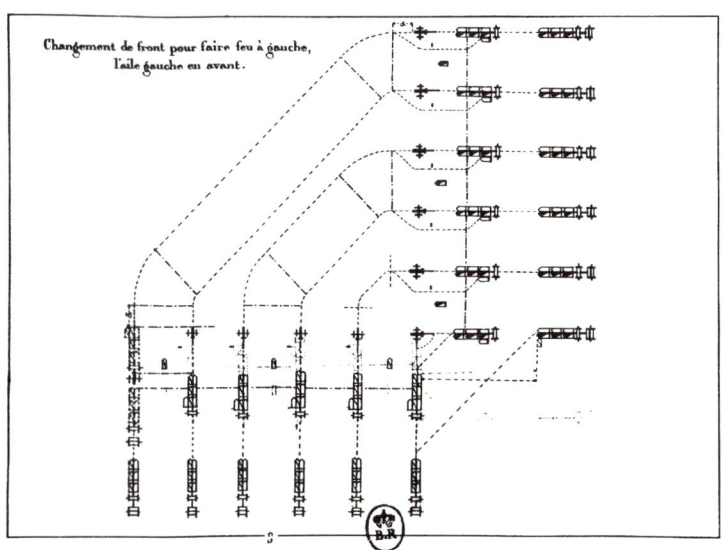

Changement de front pour faire feu à gauche, l'aile gauche en avant.

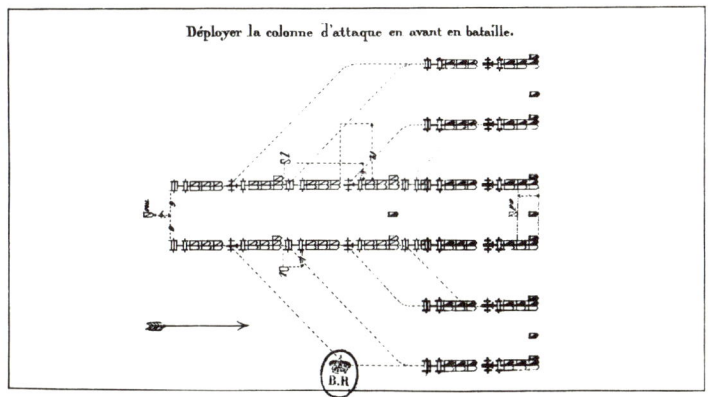

Déployer la colonne d'attaque en avant en bataille.

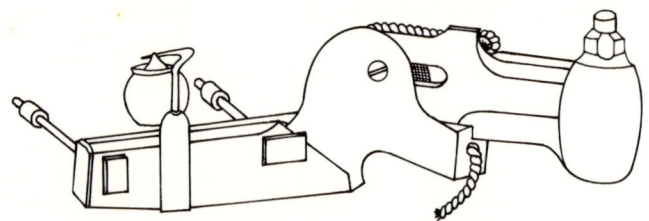

A percussion-lock for a carronade, which was placed over the touch-hole of the gun. The lanyard was given a sharp pull, and the hammer (right) fell smartly on the round explosive cap.

of mortars and howitzers, after attempts to discharge them from cannon had disclosed a tendency for the projectiles to break-up and explode in the bore of the gun. However, as an emergency measure during the siege of Gibraltar in 1779-83, the British defenders successfully discharged short-fuzed mortar shells from their 24-pounder cannon (p. 78). And about the same time, Lt Henry Shrapnel independently began the experimentation which resulted in the gun projectile which he termed 'spherical case', though some two decades were to pass before it gained the attention of a hitherto unenthusiastic British Government. As originally conceived, the shell contained musket balls intermingled with powder. But when it was found that friction was liable to cause premature ignition, the two components were separated by a thin metal partition, so giving rise to the aptly-named diaphragm shell.

As shrapnel, and what came to be known as common (powder-filled) shell, found increasing use in

guns, the problem of devising a reliable means of exploding these missiles over a designated target area became a matter of prime importance. The early timing devices consisted merely of a piece of impregnated cord which burned at a known rate, so that the moment of explosion was determined by the length of the fuze—if all went well. By the middle of the century, a variety of such fuzes were in use, and in 1849 the improved version of Captain E.M. Boxer made its debut. This comprised a powder-filled plug (for insertion in the shell), pierced by a number of holes. Firing the gun ignited the fuze's central core of powder, which then burned down to a pre-selected gap, through which the flame passed and so reached the bursting charge of the shell.

With the advent of the percussion cap, shells which exploded on impact were devised, as were more sophisticated timing devices. In one such design, employing a technique known as set back, the forward movement of the missile in the barrel brought the percussion cap into contact with a free-floating striker, thereby causing the ignition of a powder train. But experience showed that the same sequence of events could also be set in motion by an accidental dropping of the shell, a highly disconcerting discovery which led to the incorporation of a safety device (based on the rotation of the missile) which rendered it bore safe.

These developments were accompanied by attempts to increase the strength of guns without there-

The percussion-tube, an English device, produced an explosion after being struck with a hammer mounted over the touch-hole. Right: a shrapnel shell, showing fuze and sabot.

In the same fashion as small-arms, Naval guns were sometimes fitted with large flint-lock mechanism over the touch-hole.

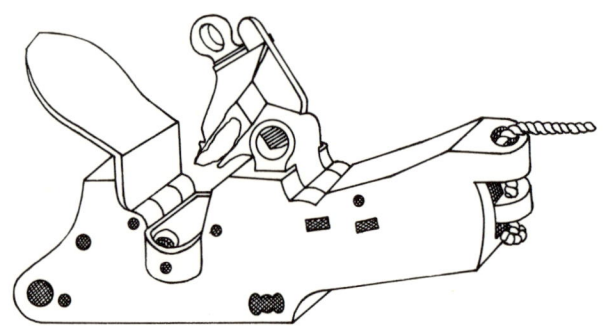

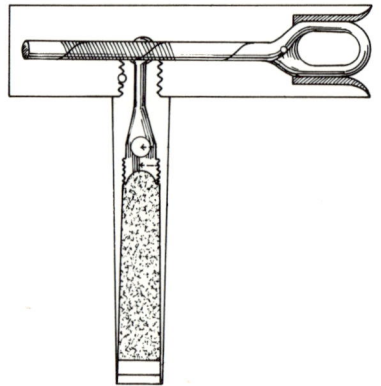

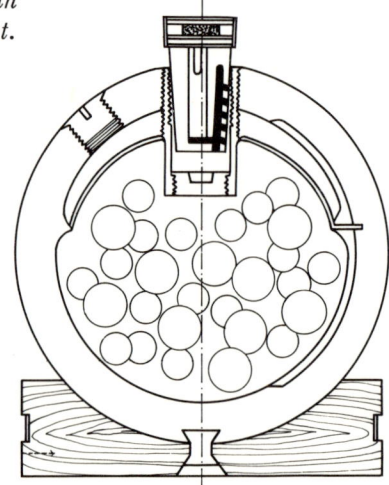

by adding to their weight. The choice of structural materials was still limited to the traditional cast iron or bronze, of which the last named, though expensive, offered the advantage of relative lightness, coupled with the fact that it was non-rusting. Bronze, however, was an alloy which was not particularly hard wearing, and as a metallic mixture it seemed to offer little scope for further improvement by means of a change in composition. This conclusion, as it happened, was not a valid one, but by the time (1870) the beneficial effects of the addition of small amounts of phosphorus came to be appreciated, the day of the bronze gun was over. On the other hand, it had long been known that the properties of iron varied widely in accordance with the amount of carbon it contained, the end product ranging from the comparatively pure wrought iron (about 0.1 per cent carbon) to a heavily impregnated (up to 5 per cent) grade of cast iron, by way of the intermediate steels, though until the 1850s these last were destined to remain a scarce and costly commodity.

In the past, irrespective of the metal used, cannon design and construction had largely been a matter of guesswork, an art which had progressed, through a process of trial and error, with little understanding of the principles involved. Strength was equated with thickness of material, so that attempts to avoid unnecessary weight eventually led to an investigation of the bore pressures which a gun of a particular calibre was called upon to withstand. The American George Bomford, hailed in the United States as the greatest ordnance expert of his time (early 1800s), and designer of that country's massive 'Columbiad' coastal defence weapons (of 10-inch and 15-inch bore), obtained the required information by inserting pistol barrels along the length of cannon under test, and measuring the velocity with which bullets were ejected when the weapon was fired.

The knowledge so gained, coupled with an eventual appreciation of the fact that it was the metal in the immediate vicinity of the bore which was required to withstand the brunt of the enormous pressures generated in the firing chamber, led to the realisation that in a cast gun there was soon reached a limit of barrel thickness beyond which any additional

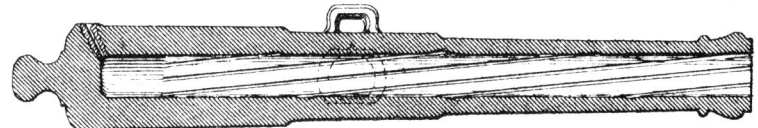

The first rifled guns were still muzzle-loaders, but the shape of the shot was changed to engage in the grooves.

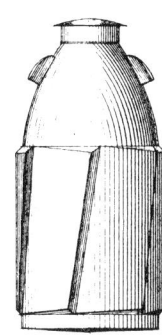
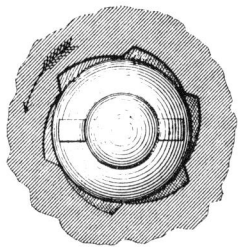

New shells in cylindrical-ogival shape. The sides are shaped to engage closely with the lands of the barrel.

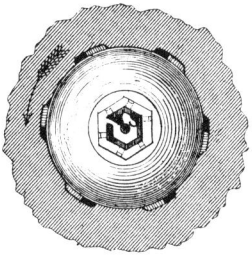
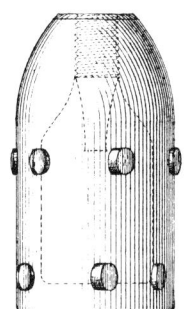

The shells have lateral studs which engaged in the rifling of the barrel.

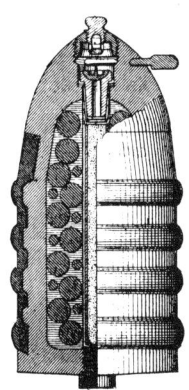
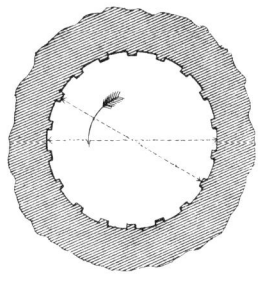

With the introduction of breech-loading guns, shells appeared with driving-bands.

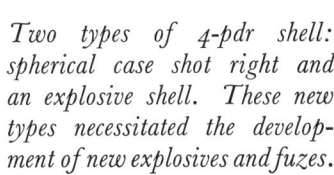
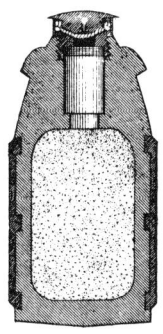
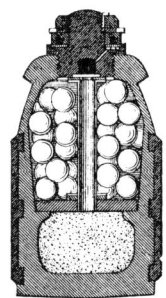

Two types of 4-pdr shell: spherical case shot right and an explosive shell. These new types necessitated the development of new explosives and fuzes.

material served little or no useful purpose in resisting circumferential stress. This important aspect of cannon design, which identified lightness and strength with the method of construction, found an early advocate in America's Captain Thomas Jackson Rodman. Like Colonel Bomford before him, Rodman was an officer associated with the U.S. Ordnance Department, and his work was influenced by the findings of the English physicist Peter Barlow, who had deduced that, in a cylinder of metal (such as a gun barrel), the pressure exerted by its different parts varied inversely as the square of the distance of those parts from the axis. Rodman's response was to introduce a mode of construction which put the metal surrounding the bore under a permanent compressive strain, and his guns were thereby enabled to withstand considerably higher working pressures than would otherwise have been the case.

To achieve his aim, the inventor reverted to the original method of gun casting, in which the weapon was formed hollow by means of a core. This procedure had been abandoned a century or so earlier, after Jean Maritz, a Swiss, had shown that greater accuracy was to be obtained from a solid casting which was afterwards drilled out, thereby ensuring perfect alignment of the bore. In the Rodman process, developed in the 1840s, an essential feature was the use of a removable hollow core, through which a stream of water passed during casting. As a result of this cooling, successive layers of metal, from the bore outwards, were compressed by the shrinkage of overlying material, thus giving rise to a weapon of increased strength and greater durability. The American authorities, however, were not impressed by Rodman's proposals, and at first declined to assist in their promotion, let alone adopt them, and the task of development was undertaken without governmental aid. Official acknowledgement and acceptance came later, by which time cast guns, at any rate in Europe, were on the point of being superseded by built-up weapons of wrought iron and steel.

Although, as yet, the major powers had managed not to engage in open warfare among themselves, this is not to say that armed conflict was altogether avoided. Russia, with the threat of a Polish revolt on her hands, had not been averse to assisting the Greeks in their struggle to gain independence from the Turks; the Austrian overlords of Italy likewise found their authority challenged; the British gradually extended their territories in India, Africa, and elsewhere; while in the Americas, an expansionist United States laid claim to Texas, and then annexed the whole of California, Arizona, and New Mexico.

And end to this comparatively tranquil state of affairs was heralded by the impending collapse of the domain of the Turks, not to mention the nationalistic aspirations of Italy and Germany—the one an open invitation to territorial acquisition, the other a potential threat to the dominant position of France. Nor was the situation improved as, one by one, the architects of the Vienna settlement vanished from the scene, to be replaced by new leaders who did not share their respect for the *status quo*. Irreconcilable clashes of interest among the great powers thus became inevitable, the first of which was not long delayed. In 1853, Russian forces occupied the Danubian principalities of the crumbling Ottoman Empire, and although Tsar Nicholas I professed to have no designs upon Constantinople, he made it clear that he was not prepared to allow that city to fall into any hands other than his own. Loud expressions of Anglo-French disapproval of the Russian move encouraged the victims of the aggression to declare war on the invaders, whereupon the fleets of the two protecting powers entered the Black Sea, there to join forces with the aggrieved Turks, whose fleet was sunk by the Russians at Sinope.

The Crimean confrontation which then ensued can be dismissed, not merely as militarily insignificant, but as providing a classic example of how not to wage war. On the allied side, the conduct of affairs was characterised by geographical ignorance, by insufficient preparation, by inadequate supplies, by a lack of reserves, by ill-drafted orders, and by ineffectual leadership. Britain's Field Marshal Lord Raglan was 65 years of age and a veteran of the struggle against Napoleon and constantly referred to the Russian enemy as "the French", while his troops, not inappropriately, were armed with cannon, some of which were reputed to have been in action at Waterloo. Fortunately for the intruders, Russia's General Prince Alexander Menshikof showed himself to be hardly less inept a commander than his opponents: with 80,000 men at his disposal, he made no move to oppose the enemy landing at Eupatoria, but chose instead to contest their advance southwards on the line of the Alma, from which river his troops were driven back and eventually obliged to make their stand at Sebastopol. The subsequent investment, bombardment, and capture of this fortified outpost duly followed, though not without a long and desperate struggle. In preparation for the attack, the Russian position had been

greatly strengthened under the direction of Colonel Todleben, an outstanding military engineer. In consequence, the defenders were able to hold out for the best part of a year, with the fire of their 18-pounder field guns augmented by that of 24-pounder howitzers and 32-pounder cannon.

Much of the allied artillery was of a lighter calibre (many of the field-pieces upon which they relied were 9-pounders) and when it became evident that the siege was likely to be prolonged, both the French and the English independently sought to increase the effectiveness of their fire by resorting to guns with a grooved barrel. Presumably this move

The Prussian field-artillery can perhaps be considered as the last of the smooth-bore cannon. The gun retains the elevating-screw on the breech, and the foresight on the muzzle, that were introduced in the preceding century. It has a touch-hole, but the charges were ready-made cartridges.

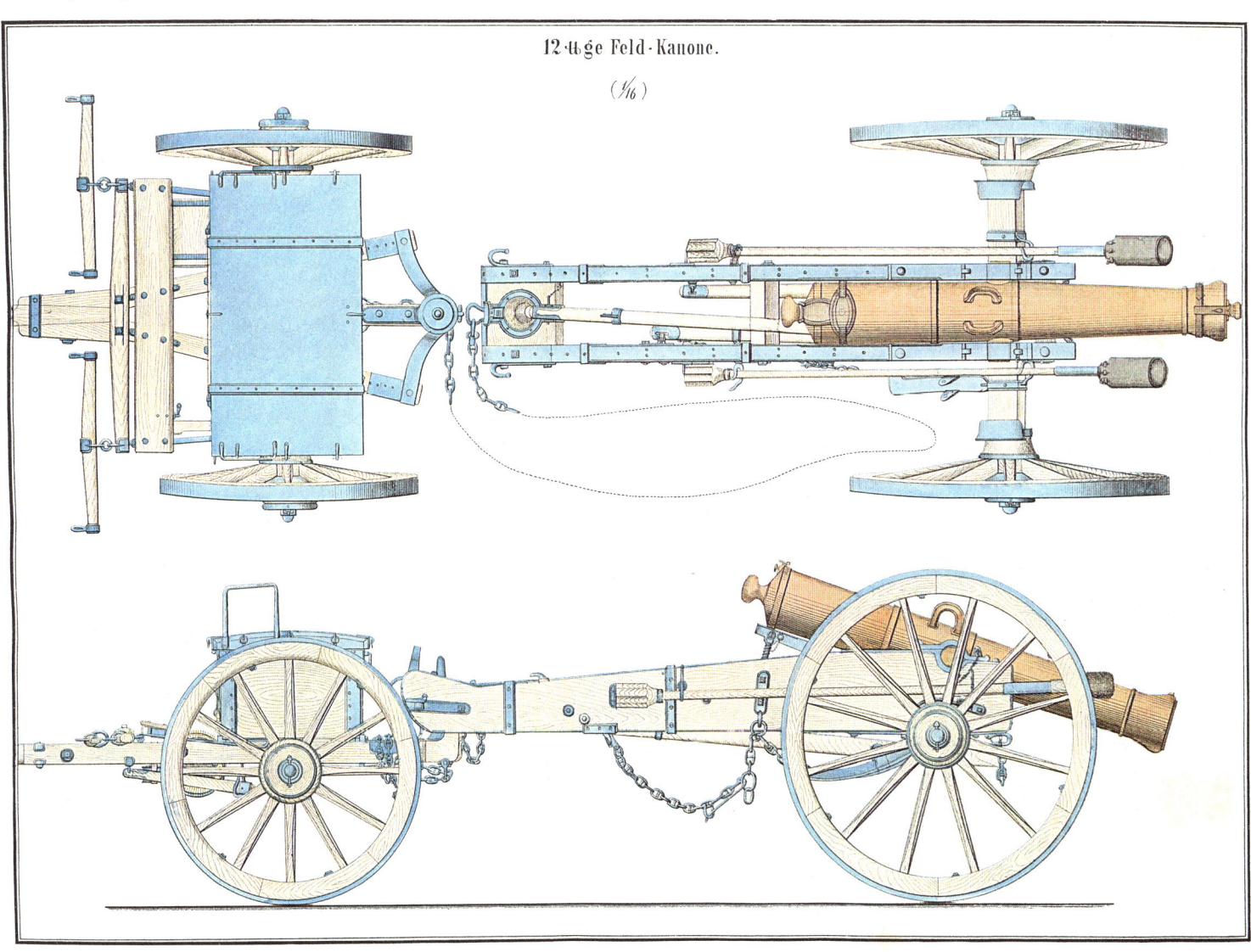

12·tbge Feld·Kanone.

(⅟₁₆)

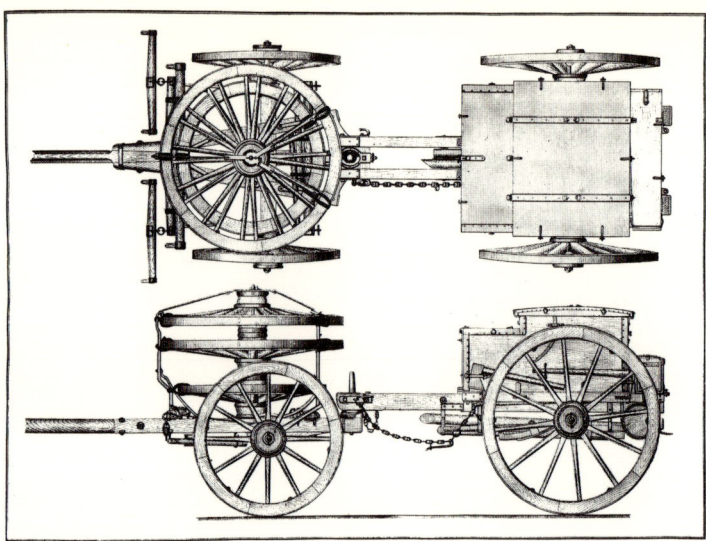

The Prussian artillery, in common with the other European powers, standardised all equipment—limbers, magazines, wheels and shafts—so that parts were interchangeable.

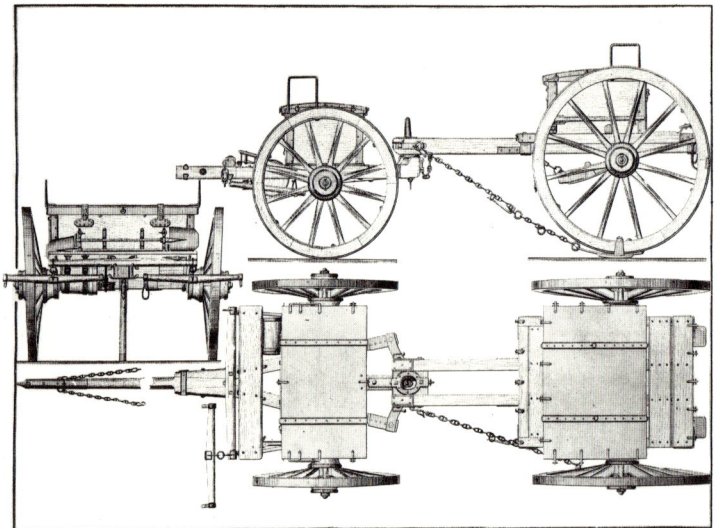

Ammunition waggons were specially made to contain various types of charges and shot. The coupling between the limber and waggon was extremely simple and easy to connect.

The field-forge carried all the necessary equipment—anvil, bellows, furnace and tools were all mounted; the farrier-sergeant also stocked shoeing material for the battery horses.

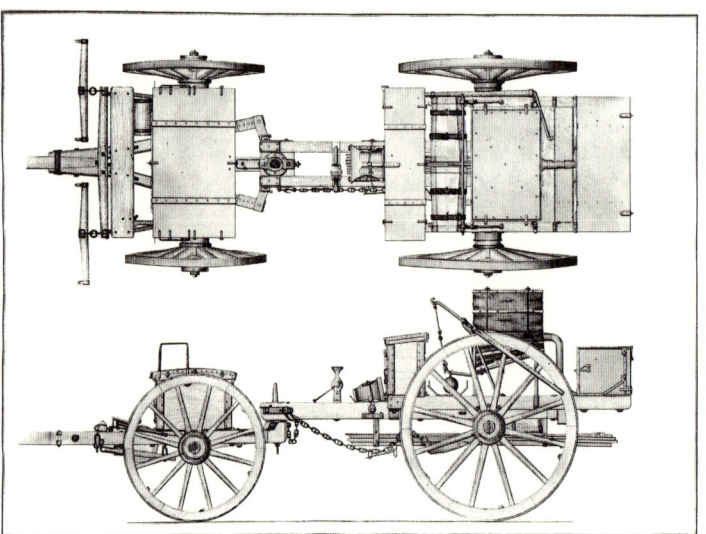

was inspired by the fact that during the earlier part of the campaign, the advantages of rifling had been brought forcibly to notice by the evident superiority of the shoulder arm carried by the French infantry. This was the new Minié rifle, sighted (albeit somewhat optimistically) at the then phenomenal distance of 1,000 yds—more than ten times the effective range of the smooth-bore musket with which many of the British troops still found themselves equipped.

The grooved interior of the Minié barrel was by no means a new idea—it was one which had found commendatory mention in Sir Hugh Plat's *The Secrets of Art and Nature*, published as long ago as 1594, at which time its application to small arms had already been attempted (Kollner, 1498; Kotter, 1520; Danner, 1552; *et al*). As time went on, rifled hand-guns found increasing favour with wealthy sportsmen and other discerning marksmen, whose continued demand encouraged its development. But by military authorities accustomed to look upon the musket as a weapon whose usefulness was limited to close order volley firing, in which accuracy of aim was a secondary consideration. The rifle was for long regarded as far too costly a weapon to place in the hands of the common soldier. In the event, it was the French who were the first to see the light, though they did not do so until 1848.

The application of rifling to cannon was a different matter, in that its promotion was hardly likely to become the concern of non-professional users. Even so, it was not without influential advocates who both appreciated and proclaimed the advantages it had to offer. Thus Benjamin Robins, in his *New Principles of Gunnery* (1742), predicted that:

"Whatever State shall thoroughly comprehend the nature and advantage of rifled barrel pieces, and having facilitated and completed their construction shall introduce into their armies their general use, with a dexterity in the management of them, will by this means acquire a superiority which will almost equal anything that has been done at any time by the particular excellence of any one kind of arms."

These were prophetic words, which did not remain unfulfilled during the next hundred years for the want of trying. The immediate difficulty lay not so much in the grooving of the gun itself as in the missile it was intended to fire, a problem which, in the case of small arms, proved to be far less acute: a slightly oversize lead ball was hammered down the muzzle, and so forced to engage the rifling on its way out of the barrel. But this was at best a makeshift procedure, and one that was hardly applicable to a weapon discharging relatively large shot made of cast iron.

A possible solution was put forward in 1790 by Joseph Manton, a well-known English gunsmith. Manton had devised a rifling machine, with the aid of which he produced a bronze cannon whose barrel displayed as many as 16 grooves. From it he proposed to fire round shot equipped with a tight-fitting softwood base, cupped to take the missile. The Master General of the Ordnance, in the person of the Duke of Richmond, was sufficiently impressed by the idea to arrange for a test. This took place at Goodwood, the Duke's country seat, between what was officially described as 'Rifled and Plane Medium 6-prs.' But, disappointingly for the inventor, the two guns did not differ greatly in their performance, and the authorities were provided with no incentive to adopt either rifling or Manton's wooden sabot.

What was not realised at the time was that rifling could in any event do little to increase the effectiveness of the traditional cannon ball, and that what was required was an elongated missile. Neither, it seemed, was there yet appreciation of the fact that the converse was equally true—that elongated missiles were impracticable in the absence of rifling. This shortcoming (though seemingly not the reason for it) had already been noted in the course of a series of experiments undertaken by the British Admiralty with the aid of naval 12-pounders, in which a means was sought whereby the weight of shot could be increased without the necessity of resorting to a larger calibre of gun. The use of two round shot was tried, but except at close range, the twin missiles diverged in flight to such an extent as to render aiming unpredictable. Nor, apparently, was this difficulty satisfactorily to be overcome by substituting a single shot of equivalent weight, in the guise of a cylinder with rounded extremities. In its

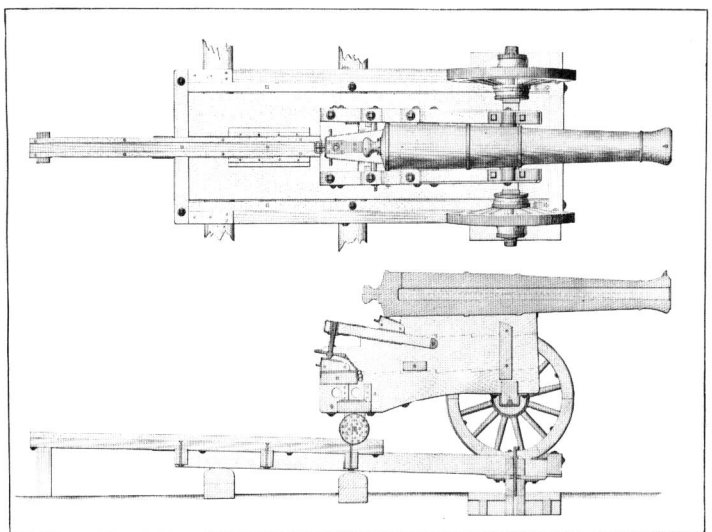

This 24-pdr fortress gun has a modified elevating screw to carry the heavy barrel, and a mounting to absorb some of the recoil shock, always a problem with heavy guns on stone platforms.

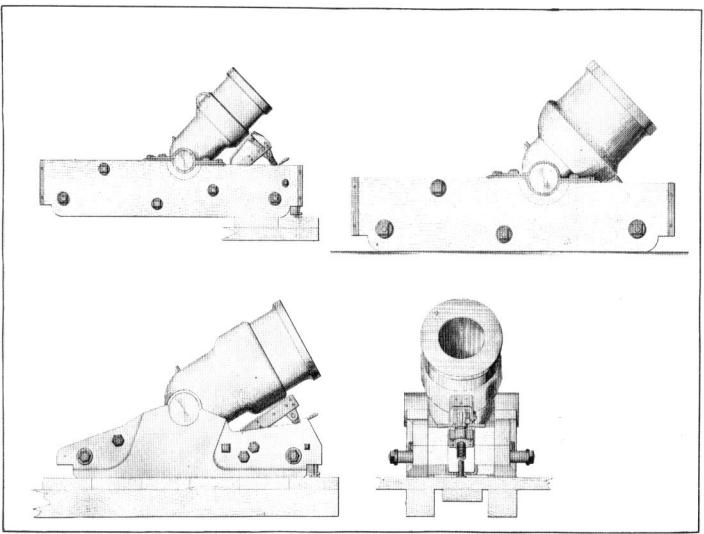

50-pdr mortars were fortress pieces or siege weapons. They had either wooden or iron mountings, and the elevating screw was found in front, before the trunnions.

This 24-pdr siege gun has a similar carriage to the field guns, but for transport, the barrel is lifted back along the carriage, so that the heavy weight could bear more equally on a simple limber.

25 ℔ge Haubitz-Laffete zur Festungs-und Belagerungs-Artillerie.

(Constr. v. 1832)

(⅟₁₆)

Mittlerer Längendurchschnitt.

Anmerkung

Die 25 ℔igen Haubitzen welche nur zum Gebrauche in den Festungen bestimmt sind, werden mit niederen Rädern und Wallprotzen transportirt, dagegen die zur Belagerungs-Artillerie mit 12 u. 24 ℔igen Belagerungslaffeten Rädern u. Protzen.

Zum Transport als Belagerungs-Laffete.

Belagerungsprotze.

Die Glacistrale beträgt 5050.

A heavy, 25-pdr howitzer, called a siege or fortress piece, was a short, fat gun. A massive carriage absorbed part of the recoil.

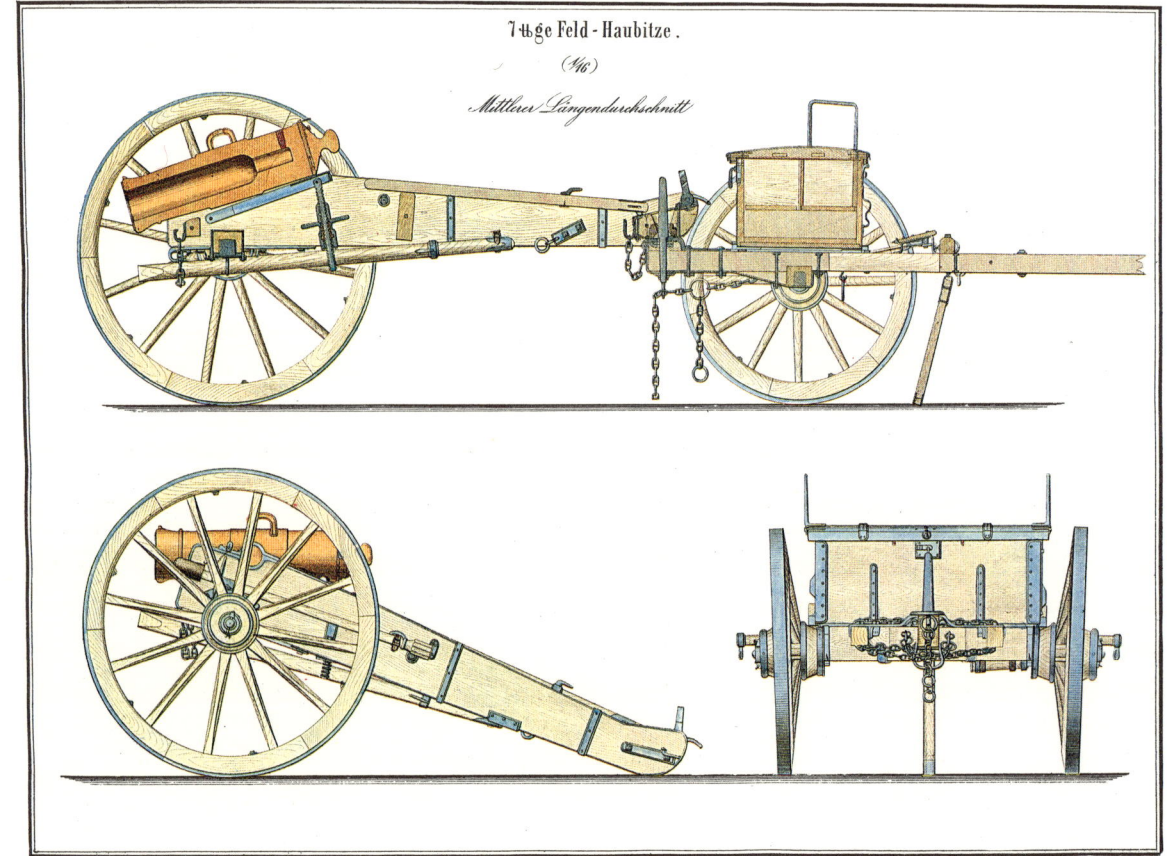

7 ℔ge Feld-Haubitze.

(⅟₁₆)

Mittlerer Längendurchschnitt.

A 7-pdr field-howitzer. It has a cylindrical powder chamber, and its elevating gear and carriage are similar to that of cannon.

136

passage through the air, such a bolt displayed a disturbing tendency to tumble end over end, also to the detriment of accurate shooting and penetration.

Thus the use of single round shot continued, not for the reason that it was in any way ideal, but because it was the only form of missile suited to the smooth-bore gun. But this over-riding consideration apart, the sphere was in fact a shape which had little to recommend it, and by its retention the performance of cannon was adversely affected in terms of firepower, accuracy, and range. One cause of this poor showing was that the diameter of the (often far from uniform) shot was necessarily less than that of the barrel, and that this difference (windage) provided a gap which permitted a considerable wastage of propelling power. To make matters worse, the escaping gases caused erosion of the bore and vent, thereby constantly adding to the loss of efficiency. Moreover, and once again thanks to windage, the movement of the shot in the gun took the form of a series of bounds, as it bounced, first off one part of the interior of the barrel, and then off another, so that the precise direction which the ball took was governed by chance, in that it depended on where it made its last contact before leaving the muzzle. Again, at moderate (i.e. then prevailing) velocities, air resistance varied directly with the square of the diameter of the missile, and only indirectly with its weight. Here, then, was yet another reason why round shot was not to be considered an ideal shape, as any attempt to add to its weight automatically entailed an undesirable increase in its diameter.

This was a disadvantage which clearly did not apply to an elongated projectile, so that, on the face of it, all that was necessary to solve many of the problems of the artillerist was for propounders of the rifled gun to adopt the cylindro-ogival bolt, that the grooves of the one, by the impartation of spin, might ensure the stability of flight of the other by maintaining the missile's axis along the tangent to the trajectory, thus ensuring that it continued on its way point first. Viewed in retrospect, such a solution appears at once simple and obvious—and it was one, indeed, which had found an advocate in Benjamin Robins as early as 1745. But at this time there were

engineering and other difficulties in the way, and almost a century needed to pass before this long awaited association of ideas eventually came about.

Progress began to be made from about 1836 onwards, heralded by the efforts of Joseph Montigny, a Belgian, and those of his countryman, General C.F.T. Timmerhans. Montigny, of *mitrailleuse* fame, after failing to obtain the support of his own Government, managed to gain the attention of the Russian authorities, by whom his suggestions, after exhaustive tests, were also rejected. In the meantime, an interest in rifled cannon was being shown in France, as evidenced by the experiments of C.L. Ponchara and the arguments in its favour put forward by Paixhans, an interest which culminated in the proposals of Colonel Treuille de Beaulieu. His idea was to endow an elongated missile with spin by means of projecting metal studs, suitably spaced so as to engage a matching set of spiral grooves in the barrel.

This plan was adopted by the French in 1842, and the idea was taken up by other countries, Austria, Russia, and Great Britain among them. At the same time, a host of alternative systems began to appear, exhibiting variations, not only in missile design, but also in the nature of the rifling. As to this, the provision of at least two grooves was plainly a minimum requirement, but the upper limit was extended to a score or more, displaying dimensions which ranged from shallow to deep, and from narrow to wide. Moreover, the amount of twist, as determined by the inclination of the grooves to the direction of the bore, could be either uniform or increasing, concerning the relative merits of which much argument subsequently arose. Unlike Great Britain, the majority of Continental powers came to favour a parabolic development for their field-guns, in which the amount of twist increased muzzlewards from one turn in fifty to one in twenty-five. Nor was this the only alternative which remained open to designers, for they were also free to arrange for the spin of the projectile to be to the left or to the right. But whichever direction of rotation was chosen, it was found to be the cause of deviation (drift) in flight, so that it became necessary to incorporate the appropriate correction in the gun sights.

At best, the use of studded projectiles was a clumsy innovation, and a number of other proposals soon came to be tried. Giovanni Cavalli, of Sardinia, produced a rifled gun which fired a ribbed shell, as did Baron Martin von Wahrendorff, of Sweden, who also experimented with lead-coated missiles. Elsewhere, the manifest difficulty of adapting shot and shell to grooved barrels inspired a certain John Mackintosh to patent a means of endowing a projectile with spin without resort to rifling at all: his idea was simply to mount a smooth-bore gun in a frame, wherein it could be rapidly rotated immediately prior to discharge...

Yet another approach to the problem likewise involved the use of a smooth-bore barrel, albeit one of oval section. The notion appears to have originated in Germany in 1696, and as developed a century and a half later by Charles Lancaster, a London gunsmith, it reappeared in the form of a hand-gun with a bore of elliptical shape which twisted upon itself. The idea attracted the attention of the British authorities, at whose instigation it was applied to both military rifles and cannon. In 1854, a number of smooth-bore weapons (68-pounders, and guns of lesser calibre) were converted to the Lancaster principle, and used in the bombardment of Sebastopol, an exercise they shared with a pair of experimental French field-guns. These last were rifled on the de Beaulieu system, and by all accounts functioned well. The British guns, however, were less satisfactory, several of them bursting when their oval-shaped, elongated missiles became wedged in the bore.

So-called rifling of a still more unorthodox nature was patented in 1855 by Joseph Whitworth. It consisted of a non-circular bore along lines earlier proposed by the engineer M. I. Brunel, father of Isambard Kingdom Brunel of Great Western fame, who had advocated an octogonal section on the grounds that a similarly shaped missile "would centre itself, both in position and in direction, to the axis of the barrel by the peculiar action of a polygon within a polygon acted upon by an increasing pitch". Whitworth adopted six sides instead of eight, and his design was applied to both small arms and cannon, at first with promising results. In 1862, a hexagon-shaped shell, discharged by 27 lbs of powder, penetrated a thick metal plate at a range of 800 yds. But despite this encouraging showing, Whitworth guns were found to suffer from a serious defect which precluded their adoption for Service use: their angular bore soon accumulated an excessive amount of fouling which no amount of scraping could readily remove.

By this time, whatever the final form it was destined to assume, it was clear that rifling of ordnance had come to stay, despite the not inconsiderable complications to which its adoption gave rise. Thus, by reason of their increased weight, elongated missiles required either a heavier charge, or the use of a more powerful propellant; and this, in turn, aggravated the problem of recoil and at the same time emphasised the need for stronger guns. Ironically enough, although the black powder of old still remained the only known explosive suitable for use in fire-arms, much more powerful alternatives had recently been discovered, only to prove unusable in their then form.

In 1833 Henri Braconnot, a French chemist, obtained a highly inflammable substance by treating sawdust and cotton with nitric acid, as did T. J. Pelouze a few years later, and in 1846 the discovery of a new "cotton explosif" (independently made about the same time by F. J. Otto and R. C. Böttger) was announced by C. F. Schönbein at a meeting of the *Naturforschende Gesellschaft*, held in Basle. Nor was this all, for almost simultaneously with the news of the advent of gun-cotton, word came from Italy of Ascanio Sobrero's production of nitroglycerine.

The high hopes aroused by these discoveries were quickly dashed when both substances proved to be unmanageable. Disastrous explosions attended their large scale manufacture, which was perforce abandoned, at all events for the time being. But their great strength was not the only inducement which ensured the continuation of efforts to find some way of bringing them under control, for the new explosives proved to be comparatively smokeless. By contrast, rather more than half of the products of combustion of ordinary black powder were solids, whose particles fouled the gun barrel and gave rise to an attendant smoke cloud. On the field of battle, such a cloud at once revealed a battery's position and tended to obscure its target. In the event, by way of the non-gelatinous (and excessively fast-burning) powders developed by Captain Johann F. E. Schultze and Walter Reid, it was not until the 1880s that P. M. E. Vieille succeeded in plasticising nitrocellulose with a mixture of ether and alcohol, so rendering it suitable for use in firearms, and that James Dewar and F. A. Abel performed a like service for nitroglycerine by combining it with guncotton to form cordite.

Long before this, the problem of providing a propellant suitable for use with elongated missiles had been solved with the aid of existing materials. Such an approach was not without precedent. Advantageous changes in the composition of gunpowder (at one time an ineffective mixture of equal parts of all three constituents) were followed, early in the sixteenth century, by an improvement in its burning rate, thanks to the introduction of a corning process, whereby the then prevailing fine (or meal) powder was converted into coarse grains, thus promoting speedy and uniform ignition.

Having regard for the increased weight of the new projectiles, coupled with the resistance offered to their travel along the length of the barrel by the rifling, what was needed was the application of a steady and sustained push, rather than the administration of a sudden blow. One suggested method of achieving this desideratum was to retard combustion of the powder by using partly, instead of completely, charred wood in the mixture. Another idea, widely adopted, was that devised by Rodman, in the United States. He carried the corning process a stage further, by producing a propellant in the form of large hexagonal prisms, each with a round hole in its centre. Unlike small-grained powder, which when ignited provided a large volume of gas almost instantaneously, the large grains were consumed relatively slowly, and at a rate which was governed not only by their size, but also depended on whether they were made to burn from the outside inwards, the inside outwards, or in both directions at once.

As for the problem of what an inventor by the name of Henry Edward Flynn described as 'the evil effects of the recoil of cannon', these promised to become magnified to such an extent as to make outsize guns dangerous to handle. In the past, what was then a lesser nuisance had been mitigated in various ways—where guns occupied fixed positions, by resort to upward sloping runways; where mobile artillery was concerned, by means of chains, wedges, and brake blocks. None of these expedients had done more than alleviate the problem, and efforts to solve it continued to be made. Thus in a heavy cannon produced in England about 1816, the muzzle end of the barrel was pierced by a series of holes, through which some of the expanding gases behind the shot could make their escape, so helping to counter the backward movement of the piece. Serious loss of velocity and range led to the abandonment of this ingenious idea which, however, was successfully revived (in modified form) during World War II.

Yet another solution, put forward by Vincent Wanostrocht in 1854, envisaged the use of a swivel gun with a difference, the difference being that the weapon was to be a double-barrelled one, with muzzles which pointed in opposite directions! According to theory, when one of the two barrels was fired, the recoil would turn it about, leaving its still loaded companion pointing towards the enemy. No less original was a proposal made in the following year by the previously-mentioned Henry Flynn, who suggested that cannon be suspended in a frame, in such a manner that any rearward movement served to raise them against their own weight. A decade or so later, the essentials of this idea were incorporated in the disappearing carriage associated with the name of Colonel A. Moncrieff. As applied to coastal defence ordnance, the energy of recoil was utilised to lower the gun to a protected position, and thereafter to raise it when it was about to be fired. The system, in which counterweights were employed, was not without its drawbacks and in any case had a limited application which did not include field-guns. As late as the time of the Franco-Prussian War, these continued to run back on

During the summer of 1849, a French General, Oudinot, laid siege to Rome in an effort to defeat the occupying Risorgimentist patriots and to return the reactionary Papa Nono (Pius IX). The French expeditionary force had 24 siege guns.

discharge, making accurate aiming difficult and rapid firing impossible. It was not until 1888 that Konrad Haussner finally solved this long standing and vexatious problem, by interposing a recoil system between gun and carriage.

Unexpectedly enough, the seemingly much more formidable question of how to increase the strength of guns (that they might withstand the force of more vigorous propellants) and at the same time reduce their weight (in the interest of increased mobility) proved less difficult to solve, thanks to the introduction of new methods of construction and the timely discovery of how to mass produce steel. As to this, by virtue of the fact that the various steels occupied an intermediate position between wrought iron (with its low carbon content) and cast iron (with its high carbon content), theory suggested that it should be possible to accomplish an iron-to-steel transformation, either by carbonising the one, or by decarbonising the other. The Ancients, who did

not have access to cast iron (since they did not possess a furnace sufficiently hot to melt the metal), perforce adopted the slow and costly procedure of packing wrought iron with charcoal, and heating it for a week or more, in the course of which carbon was gradually absorbed. The product, which, because of its appearance, came to be known as blister steel, varied greatly in texture and composition until the advent, in 1740, of Benjamin Huntsman's crucible (or cast) steel. This was a much superior fused product of uniform quality whose manufacture remained a British monopoly for the next 100 years. Then William Kelly, an American, demonstrated that molten pig iron could be decarbonised merely by blowing air through it, and in 1855 Henry Bessemer devised a converter for carrying out the process, and subsequently acquired the Kelly patents. Bessemer's converter revolutionised production, in that it enabled steel to be made in large quantities in very short time and at

140

low cost. Almost overnight, the price of the material dropped to less than a quarter of its former value. The day of the all-steel gun, however, still lay in the future, and it was preceded by an intermediate period during which, though the casting of weapons still continued, this process was gradually replaced by a new technique which offered the advantages of increased strength combined with lightweight construction.

The modern concept of the built-up gun dates back to the 1830s, and has been attributed to the American engineer Daniel Treadwell, though credit for the idea has also been claimed by a Frenchman (Thiery). However this may be, Treadwell, Rumford Professor at the University of Harvard, conceived and developed a novel form of cannon construction which he patented in 1841. In its essentials, his gun consisted of a central tube surrounded by metal hoops, shrunk on after heating. This squeezing process gave rise to tangential compression along the length of the tube, an endowment which any internal gas pressure would first need to overcome. The same effect was later obtained by other means, such as wrapping the barrel with wire under tension (Longridge, 1855) and the application of hydraulic pressure (Whitworth, 1862), not to mention the casting process which Rodman introduced in 1845. But although Treadwell hopefully founded the Steel Cannon Company with the intention of marketing his invention, the enterprise did not flourish, and in the years which followed he had the mortification of watching the exploitation of his brain child by others. Among his emulators was a compatriot by the name of Robert Parker Parrott who, on patenting his own design for a strengthened gun in 1861, thereafter found himself sued for infringement. But by this time, Treadwell's cause was already lost, for the built-up gun had long since appeared in a multiplicity of forms abroad, particularly in England, where this method of construction had been developed by such advocates as Alexander T. Blakely, Lynall Thomas,

Following the example of other Italian towns in 1848, Venice attempted to throw off the Austrian yoke. The Austrians left the city, but surrounded it with guns. This battery (left) was sited on the island of San Giuliano.

and Joseph Whitworth. Above all, the notion had been taken up by William G. Armstrong, a Tyneside engineer who was soon to become recognised as its leading exponent.

It was the outbreak of war in the Crimea which brought Armstrong into the armament business, as an additional outlet for his engineering interests. He chanced to be staying in London at the time of the Battle of Inkermann, concerning which reports appeared in the Press. These described how the crews of two British 18-pounder guns had silenced an entire Russian battery, only to find that they could not exploit this success by advancing to a new position because the excessive weight of the cannon (nearly 3 tons) prevented their being moved over the rough ground. Armstrong promptly made a sketch of a compact field-gun, in effect an enlargement of the latest infantry weapon, rifled and intended to fire elongated instead of spherical shot. He submitted this design to the Secretary of State for War, by whom he was actually encouraged to develop the idea.

Armstrong's first thought was to take advantage of the superior tensile strength of steel (nearly twice that of wrought iron), but at the time (early 1855) all he could obtain was a forging suitable for a gun with a calibre of less than two ins, and this only after no less than eight barrels had been rejected because of flaws in the material. This experience convinced Armstrong that despite the advantages which steel undoubtedly had to offer, it was not yet available in the high quality required for gun making. He accordingly turned to the more dependable wrought iron, utilised in a particular manner which he described as 'forming tubes by rolling up iron bars in spiral coils and then welding longitudinally'. Pending the availability of steel barrels, the early Armstrong guns were made up of a series of concentric tubes, heated and shrunk, one upon another, and in 1859, after successful tests, a branch of the inventor's engineering concern was formed, under the name of the Elswick Ordnance Company. It was financed by, and manufactured exclusively for, the British Government, in close association with the Royal Gun Factory at Woolwich, and one of

Armstrong's first assignments was to replace the whole of England's field artillery with the new weapon. As a result, horse batteries received 9-pounders weighing six cwt and field batteries 12-pounders weighing eight cwt, and the extent of the increase in mobility which this re-equipment achieved may be judged from the following tabulation of gun weights. It relates to 9-pounder weapons:

Weight of gun		
Cast iron	Bronze	Built-up
26 cwt	13 cwt	6 cwt

The hostilities in the Crimea not only convinced the contenders of the need for rifled cannon, and so led to the widespread adoption of the built-up gun, but it also proved to be a fateful event for makers of steel. When the fighting began, Henry Bessemer approached the British War Office with an idea for a method of rotating projectiles, only to have it rejected without so much as a trial. Experiments he then carried out in France under the aegis of Napoleon III revealed the inadequacy of existing weapons when they were called upon to fire the new missile, and this set him to searching for "a superior description of cast iron" better suited to gun manufacture. It was this investigation which led to the devising of a means of mass producing steel, though once again Bessemer was unfortunate in his approach to the British Government. As luck would have it, his representations coincided with the rise to eminence in the realm of armaments of William Armstrong, the weight of whose authority was such that his forthright dismissal of the Bessemer product as 'totally unfit for the manufacture of ordnance' sufficed to prevent any attempt at its utilisation by official establishments for a number of years. The knowledge that this prohibition was based on the presence of minute gas bubbles in his steel prompted Bessemer to make strenuous efforts to eliminate the defect, in the course of which he made and tested guns himself. Among others who also shared his faith in the possibilities of the material was Joseph Whitworth, who devoted his considerable resources to the devising of a method of overcoming porosity in the metal by subjecting it, while molten, to hydraulic compression. But after years of endeavour had been spent in the perfecting of this purely mechanical process, it was superseded by chemical and other means.

The leading advocate of the steel gun in Europe, destined to become Armstrong's most formidable rival, was the Frederick Krupp concern, of Essen. Until the Napoleonic blockade of the English Channel, Germany and other Continental countries had obtained much of their cast steel from Great Britain. When supplies of this vital commodity were interrupted, attempts at home production were encouraged, an undertaking in which Frederick Krupp decided to join. In the years which followed, the high quality of Essen steel gained well deserved renown for its makers, and the manner of its development was widely attributed to Krupp genius, not to say mystique. Doubtless innate skill and ability played an important part, but the more prosaic facts of the matter suggest that the Teutonic success was not unconnected with a visit which Frederick's son Alfred made to London in the summer of 1838. At all events, it is known that, thinly disguised under the *nom-de-plume* of A. Crup, he subsequently toured the main manufacturing centres of the British steel industry, gathering what information he could. Alfred Krupp began making armaments in 1853, with the trial production of what he claimed to be "the first mild steel musket barrel ever produced", and four years later he completed a 3-pounder cast steel cannon. But although he sent this weapon to Spandau for official appraisal, two years went by before the authorities troubled to test it. The weapon was then rejected, not because of any fault in its performance, but on the grounds that existing guns were hardly in need of any improvement, especially if it entailed an increase in manufacturing costs. Krupp nevertheless persisted in his attempts to produce better guns, and at the London Exhibition of 1851 he was able to display a 6-pounder fashioned out of a single block of metal, together with a flawless ingot of cast steel which weighed 2 tons. At the Paris Exhibition of 1855, a yet larger gun was shown, but apart from a modest demand for his products which subsequently came from the Government of Egypt, it was not until 1859 that the superior merits of his all-steel cannon were acknowledged to provide a justifiable reason for increased military expenditure, whereupon orders began to pour in from Russia, Holland, Sweden, and even

During the Crimean war, French, British and Sardinians laid siege to Sebastopol. Here are some of the 609 guns of the French artillery, which altogether, during the 349 days of the siege, fired more than a million shots, of which the majority were heavier than 24-pdr shot. The action enabled trials to be made with rifled guns.

The British artillery at Sebastopol totalled 197 pieces, which fired 250,000 projectiles, mostly larger than 24-pdr. The siege mortars were photographed in front of the Picquet-House. The ready-use supply of shells lies in the foreground.

Prussia. And in 1862, Krupp turned to making steel by the Bessemer process...

In America, meanwhile, gun development was along different lines. After the Bomford 'Columbiads', which were intended to serve in the dual capacity of both howitzer and shell gun, smoothbore production was represented by the cast iron cannon of Commander John A. B. Dahlgren (designer of a series of distinctive bottle-shaped naval guns) and of Captain T. J. Rodman (who specialised in field and heavy siege pieces of up to 20-inch bore). Notwithstanding that Richard Delafield, a distinguished member of a team of observers sent to the Crimea, afterwards submitted a "Report on the Art of War in Europe in 1854-56", cast iron continued to remain the preferred material. One reason for this was the high quality of American ore, which furnished metal possessed of a tensile strength more than twice that of some of its European equivalents. Another was the recent introduction of improved methods of cannon casting (Rodman process).

Even so, there was a certain amount of experimentation entailing the use of wrought iron and steel,

and developments abroad were by no means ignored. In particular, they found an advocate in Captain Parrott, whose professional interest in ordnance was such that he ended by resigning his commission to become Superintendant of armaments production at the West Point Foundry, in Cold Springs, N. Y. In 1849, on learning of the (then supposedly secret) manufacture of rifled cannon by Alfred Krupp, he decided to adopt this feature. He was thus led to devising his patented method (to which Daniel Treadwell took such exception) of strengthening cast iron cannon with the aid of a wrought iron hoop, shrunk on the body of the gun in the region of the powder chamber. He was then faced by the problem of providing a suitable projectile for the weapon, which he sought to overcome by endowing the missile with a brass skirt. The lower part of this ring of metal was undercut in such a manner that the force exerted by the propellant gases caused it to expand into the grooves of the rifling. But as this expansion could not take place until the gun was fired, the arrangement was not altogether satisfactory. Because the shot needed to be sufficiently loose-fitting to slip down the muzzle of the piece during loading, this meant that it was free to slide forward if the elevation of the barrel was depressed below the horizontal, a disconcerting and aim-spoiling happening which could also occur if the gun was suddenly stopped in its tracks when it was being run out. No doubt this was one of the considerations which led to the rejection of the Parrott gun by the Federal authorities, though with the outbreak of civil war which soon afterwards occurred, this decision was quickly reversed. And of an estimated 7,892 cannon which were cast in Northern foundries during the period 1861-65, more than a quarter were produced by Parrott, in half a dozen or more calibres

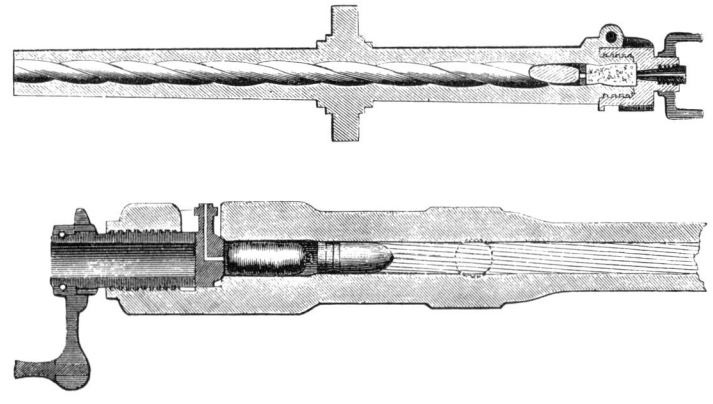

The diagrams show two types of English breech-loaders: Above, the hexagonal Whitworth gun; below, the Armstrong gun, which, although breech-loaded, still retained the touch-hole.

At Sebastopol, the Russians had 982 guns destroyed, and 3,000 carriages damaged, some being partly reparable. The power of the French artillery is shown by this photograph of no. 6 bastion, in the south of the town. It contained 40 guns.

which ranged from 10-pounders to 300-pounders. Nor did these once-despised weapons fail to give a good account of themselves. If required to do so, they could throw an elongated missile more than twice the distance a smooth-bore.

Once the principle of rifling was accepted, the American authorities were faced by the question of what to do with existing armaments, and of deciding whether these were to be (1) scrapped and replaced; (2) retained and converted; or (3) used in their existing form. But whereas Great Britain had decided on the first of these possibilities and France was content to take the middle course by adapting her bronze field-pieces to the de Beaulieu system, in the United States, though all three courses were followed to some extent, it was smooth-bore cannon which were destined to play a decisive part in the battles that lay ahead. As for rifled guns, in so far as they resulted from the modification of existing equipment, many of them proved to be a disastrous failure. In the report of an official investigation, ominously entitled "Summary of Burst Guns", it was revealed that during one naval engagement, not only had every weapon in the fleet burst, but that in the course of so doing, four times as many men (forty-five, as against eleven) had been killed or wounded by five of their own guns during the first bombardment, as were incapacitated as a result of enemy action!

The American Civil War began in April, 1861, and in terms of men, money, and materials, the outcome was a foregone conclusion from the start. The Federal authorities in the North, indeed, had an overwhelming advantage over their opponents in the South, in that they occupied twenty-three of the thirty-four States, had four times the (white) manpower to draw upon, owned two thirds of the country's railroads, exercised control over the seas, and possessed by far the greatest share of manufacturing facilities and financial resources. Yet despite the tremendous odds against them, the would-be Secessionists were able to continue their resistance in the field for nearly four years, thanks in no small measure to outstanding leadership. General Robert E. Lee, however, inevitably found himself increasingly forced on the defensive, able to delay, though not to prevent, the fulfilment of a Northern strategy which aimed at splitting his forces into two, that they might be dealt with individually and at the convenience of the attacker.

Notwithstanding Count Helmuth von Moltke's contemptuous dismissal of the struggle as "two armed mobs chasing each other round the country, from which nothing could be learned", the fighting nevertheless contained a lesson for those who chose to pay heed: an intimation that in this, the first conflict in which rifled weapons were extensively used by artillery and infantry alike, any ensuing advantage lay with the defence. Two and a half decades earlier, H. J. Paixhans had warned that in the absence of a corresponding development in cannon, the advent of rifled infantry arms was likely to reduce the role of artillery to one of secondary importance. But now, with opposing armies possessed of rifled field-guns which could shoot with greater accuracy over increased distances, it transpired that their greatly improved performance was largely nullified by the ineffectiveness of the available ammunition. For experience soon showed that shells containing such a relatively weak bursting charge as ordinary black powder offered no great threat to enemy troops, especially if (as became increasingly the case as the war went on) they had taken the precaution of digging themselves in. On the other hand, in the absence of an obliterative preliminary bombardment of a prepared position, attacking troops had to face a devastating hail of grape and cannister shot, discharged point-blank from batteries of smooth-bore weapons. Thus the American Civil War, while it demonstrated the effectiveness of artillery as used by the defence, also revealed the existence of serious limitations when it was employed as a long range instrument of assault. In the event, both sides responded by increasing the weight of their attack, but the mere massing of guns did nothing to remedy the shortcomings of the weapons themselves, and neither did the attitude of some military experts. One senior British officer actually argued that the ability of a rifled field-gun to place shell after shell within a designated target area was an undesirable attribute. All an enemy commander had to do was to keep his troops away from the region concerned, whereas if he and his men were within reach of round shot, they would not be able to consider themselves safe anywhere!

But accepting as desirable the greater range and accuracy which only a grooved barrel could provide, what was also needed was an increase in the rate of fire combined with the use of high explosive shells, and neither of these requisites was immediately forthcoming. Trinitrophenol (picric acid), which the French called melanite and the British termed lyddite, did not come to be employed as a filling for shells until the 1880s, though efforts to improve the performance of rifled cannon in other ways had continued to be made since they were first introduced.

Of these, the most important involved a return to breech-loading in 1879, abandoned in 1869 (because of obturation difficulties) as soon as the casting of heavy ordnance became feasible. But the ability to mould guns in one piece, while it solved the problem of gas escape to the rear, did nothing to prevent loss of propelling power to the front, as a muzzle-loaded projectile needed to pass easily down the whole length of the barrel. In such circumstances, the problem of windage applied to smooth-bore and rifled weapons alike, and the only satisfactory method of overcoming the difficulty appeared to be by way of a tight-fitting missile which could be inserted through an opening in the rear of the gun.

Both the Cavalli and the Wahrendorff rifled cannon were breech-loaded, the arrangement being

This and the following photograph are from the American Civil War, which marked the virtual end of a long history, when the 3 branches, infantry, artillery and cavalry, were of equal importance. The guns, protected by earthen ramparts, protected the camp.

The quick-firing gun rapidly outclassed these smoothbore bronze guns. The railway, in the Civil War, assumed a great strategic importance. For the first time, Armies and their supplies could be moved rapidly from place to place independently of horses.

that after missile and charge had been stowed in position, a cylindrical cast iron plug was inserted and held in place by a wedge, slipped into a slot cut in the body of the gun at right-angles to the bore. In addition, the Wahrendorff version incorporated a hinged iron door, the closing of which pushed the breech plug into place. Krupp in due course adopted the essentials of the procedure, upon which he improved by rounding the rear face of the slot and inclining it slightly to the axis of the gun, while the wedge was housed in a sliding breech piece whose movements were guided by metal ribs. The arrangement, which had the merit of simplicity and lent itself to speedy operation, was criticised because of the great length of breech required, with the disproportionate increase in weight this involved,

and the fact that the wedge itself promised to become an operational hazard in outsize guns.

While ribbed and studded shot did nothing to solve the question of forward windage, the introduction of breech-loading made it possible to find an answer to this age-old problem by way of missiles, part or all of which were of groove instead of bore diameter. Krupp, in his so-called compression system, used shells encircled by slots containing leaden rings, in association with rifling consisting of numerous shallow and narrow grooves, intended to provide as large a bearing as possible. But because of the extreme softness of lead, projectiles ringed by, or coated with, this metal proved unsuitable for high velocity guns, as they tended to tear their way down the barrel without gaining twist. Resort was then had to brass flanged and copper discs, designed to expand under pressure, until the ultimate answer was found in the Vavasseur driving band. As its name suggests, this consisted of a flat metal ring of ductile metal (copper) which encircled the missile near its base. While the main body of the shell rode the elevated surfaces (lands) between the grooves, the raised copper band engaged the rifling, so causing rotation and at the same time preventing forward gas leak. The associated problem of rearward obturation, which the adoption of breech-loading re-introduced, was in due course successfully countered by Broadwell (who used a combined ring and plate action), by Armstrong (the Elswick cup), and by a French Colonel who gave his name to a mushroom-headed device used in conjunction with an oiled asbestos plug (the de Bange pad). The Krupp polygroove system, meanwhile, was quickly adopted by Armstrong, who also sought to facilitate the insertion of close-fitting shot by way of the muzzle. To this end, he introduced an ingenious arrangement known as 'shunt rifling', in which a wide groove, deeper on one side than the other, permitted a studded missile to move freely down the barrel during loading. On the gun being fired, the projectile, now turning in the opposite direction, automatically moved over so that its studs engaged the shallow side of the rifling. The great strain this placed on the projections, however, was such that they often gave way, and the idea was abandoned.

Strasburg, 27 August 1870. In the afternoon, 293 cannon and mortars played on the city. The French General, Uhrich, surrendered during the 28th to avoid further civilian losses and the destruction of the city.

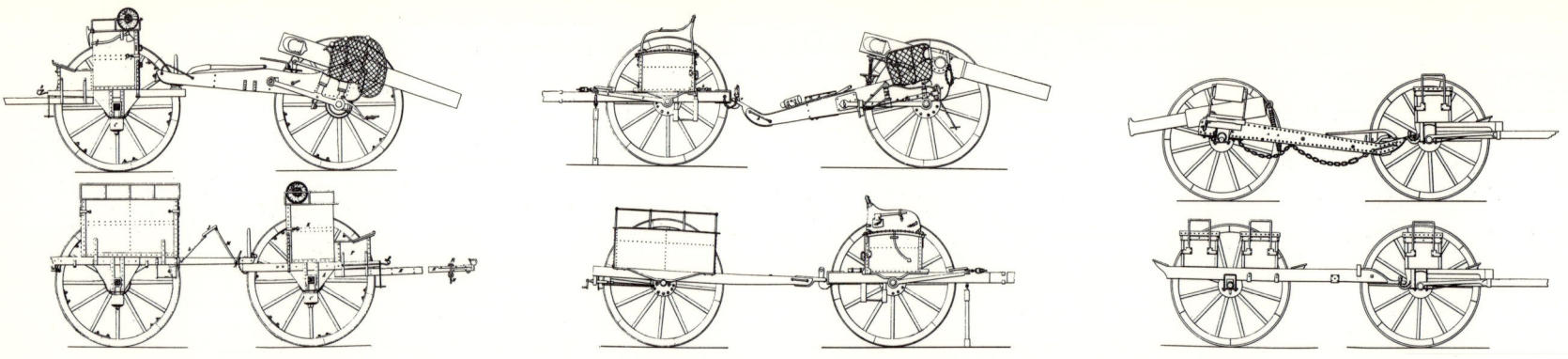

The different couplings of carriage and limber in 1870, left: the Russian method, centre: the German, and right: the English way.

So, eventually, was Armstrong's method of breech-loading, in which he closed the bore with a vent piece, held in position by a hollow breech screw, through which missile and charge could be inserted. The vent piece was then dropped into place by way of a slot, and the screw tightened. But if this mechanism did not prove to be enduring, the Krupp wedge subsequently found a more serious rival in a modified version of a screwed breech block, similar to that suggested by de Beaulieu in 1842. A decade or so later, two Americans jointly patented a design in which partly cut-away threads on both the block and the inside of the breech opening, enabled these to be firmly engaged merely by the making of a quarter turn. This idea was taken up and perfected by de Beaulieu and Verchère de Reffye, with the result that what had been a crude device was transformed into an effective and widely used method of breech closure, several variations of which (employing stepped and eccentric screws) subsequently appeared.

Before these developments occured, Great Britain had second thoughts about the desirability of breech-loading. Soon after what critics held to be the Government's precipitate adoption of the Armstrong system, a number of accidents showed that guns could be fired before the breech was properly closed.

Ironically enough, this retrograde step was taken at a time when an over-confident Napoleon III chose to match his bronze muzzle-loaders against the cast steel breech-loading guns of Prussia. The evident superiority of the Krupp armament apart, the defeat of the French was ensured by the skilful tactical handling of their artillery on the part of the German commanders. As to this, in order of march, the guns were kept well to the fore, ready for immediate use, while in action they were massed so as to provide a heavy and concentrated fire. This fire was first directed against the guns of the enemy (the so-called artillery duel), and not until these had been silenced was it then turned upon opposing troops, in preparation for an infantry attack.

The crucial encounter, which decided the outcome of the war, was at Sedan, where some 600 German cannon mounted a sustained barrage which

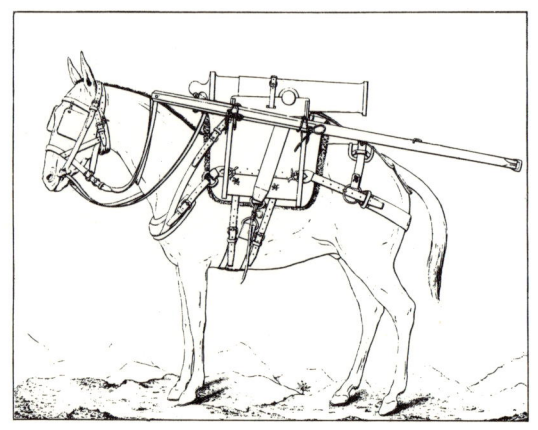

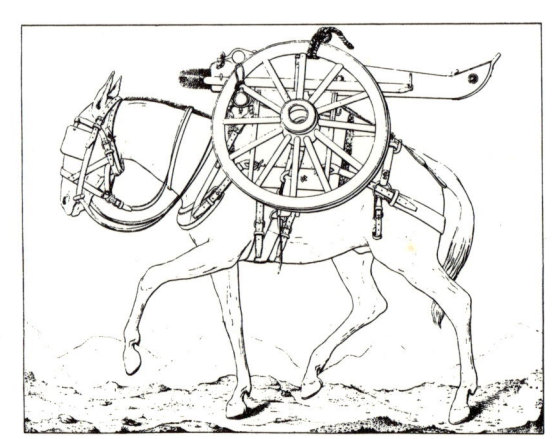

About 1830, mountain guns and carriages were lightened and made easily detachable for mule carriage. This was not new, for the British had used similar methods in 1813 in Spain.

150

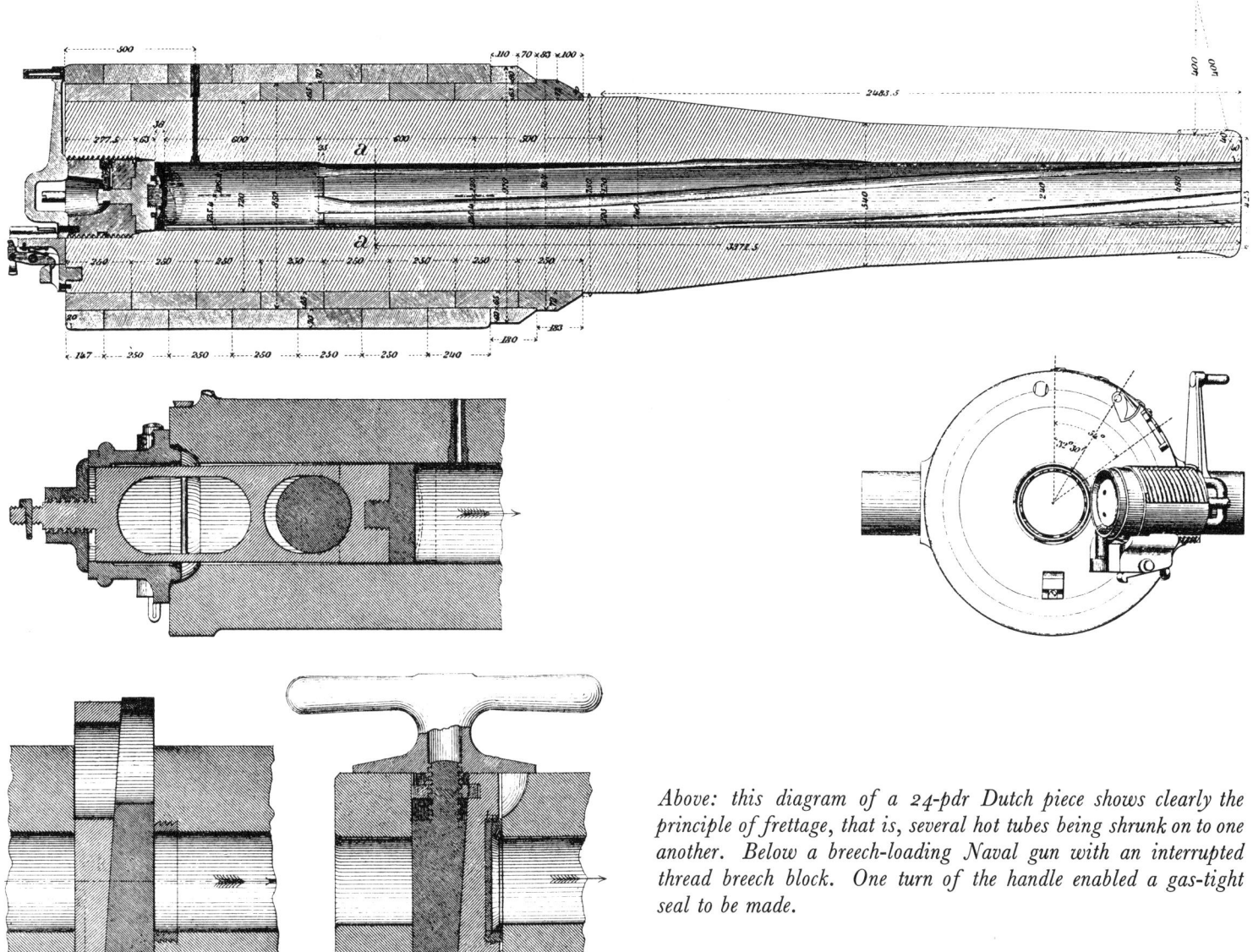

Above: this diagram of a 24-pdr Dutch piece shows clearly the principle of frettage, that is, several hot tubes being shrunk on to one another. Below a breech-loading Naval gun with an interrupted thread breech block. One turn of the handle enabled a gas-tight seal to be made.

Opposite, left: the Swedish breech mechanism of Baron Wharendorff. The seal was made by a bolt, not shown, which passed through the barrel and the sphere. Below: Kreiner's design for Prussian artillery based on the principle of two opposing wedges.

annihilated the Imperial Army. And though British observers may have derived comfort from the fact that more than 200 of the Krupp guns failed in action at one time or another because of defects in the wedge mechanism, the superiority of breech-loading rifled ordnance had nevertheless been established. But despite mounting evidence of the short-sightedness of their decision, the British authorities doggedly maintained a preference for muzzle-loading until as late as 1880, whereafter it could no longer be upheld. In the previous year, an 110-pounder naval gun had blown up, killing ten officers and men, because of an inadvertent double-

loading—a type of accident which could not have occurred, had the charge been inserted other than by way of the muzzle. But what finally led to the abandonment of the muzzle-loading gun was the use of slower burning propellants, as these entailed an increase in barrel lengths from 20 to 30 calibres and more. Thereafter, with the devising of an efficient recoil mechanism and the advent of the brass cartridge case, the way was open at last for the emergence of the rapid firing gun and the restoration of field artillery to a position of dominance in both defence and attack. From this point, modern artillery and tactics began to take their familiar shape.

The heavy artillery at Strasburg contained smoothbore mortars of 50-pdr, 25-pdr and 7-pdr types. There were some rifled 21-pdrs. This is a mortar on a siege mounting.

The German artillery bombarded Strasburg, starting on the 13 August 1870 and continuing for 40 days, causing grave losses to the inhabitants. The Prussian field artillery consisted of rifled guns, 6-, 12- and 24-pdrs, firing explosive, incendiary and caseshot shells.

THE HISTORY OF ARTILLERY IN THE LAST CENTURY
1871 - 1971

A study of modern artillery reveals that the barrel, gun-carriage, sighting, projectile, interior and exterior ballistics resulted from isolated and individual technical developments. The actual engagement of artillery in the field depends on the application of these techniques and, consequently, must be derived from them.

A study of weapons also leads to the crux of the major question of our time, man versus machine, and a certain anxiety as to where this will lead us. And so the study of artillery during the last century of its history testifies to the dramatic struggle between the often galloping rate of technological progress and the human desire to avoid being overwhelmed by it.

Although this problem is far from being peculiar to artillery, the evidence presented by this weaponry over a very long period sheds a very interesting light on our times. One extraordinary factor has been discovered by historians which, in practice, could be considered as a constant. It is revealed that the greater the pace of those isolated technological improvements, the slower the knowledge of their combat potentialities was acquired.

Over the last century during which artillery, itself, passed out of the Bronze Age, the number of patents filed by engineers relating to improvements in performance is considerably less than the total tentative efforts and miscalculations that had to be made on the battle-field where the actual knowledge of the possibilities and limits of operation is found. In other words, a familiarity with artillery enables the limits of the more exact sciences to be clearly defined and to take into account the human sciences on which they so often depend. War shows us what practice on the ranges can never do, that the human moral fibre can stand up to the highest intensity of fire under certain conditions, whilst less intensive fire skilfully placed can have surprisingly effective results. The reasons for this phenomenon are well worth examining, since this aspect has been almost totally overlooked in treatises on artillery.

THE INTRODUCTION OF QUICK FIRING GUNS FROM 1871 TO 1914

In 1871, after a long period of practically uninterrupted war, the guns fell silent and armies were reorganized, drawing on experiences which were gradually being analysed.

From a laborious compilation of statistics, lists were drawn up of losses, their causes ascribed and bad decisions by the high command, only discussed in very closed circles, were revealed. These statistics also demonstrate the role played by the different weapons, a subject which can be expanded at will as the discussion of the ideal proportions of weapons could never compromise any one person.

On this basis, it can be established that the French artillery caused about a fifth of the losses in the ranks of the German army, and that the Chassepot rifle accounted for the rest.

After its defeat in 1870, the French army sought to make up for its weaknesses, particularly in adequate field artillery. The efforts of a generation of soldiers and scientists resulted in a new system based on the famous '75'.

Conversely, statistics revealed that the actions of the German artillery caused nearly three-fifth of the French losses, the Dreyse rifle proving less proportionately effective as a killer weapon than its enemy counterpart, the Chassepot.

It is undoubtedly a somewhat artificial endeavour to make a résumé in a quasi-arithmetical form of the very complex state of affairs, represented by the quantity of different weapons in relation to the most widely varying battle situations. However, during this period, the ratio of weapons was appreciably simplified following the disbanding of the cavalry. After Reichshoffen and, of course, Balaklava, even if not overtly discussed, it was recognized that cavalry would no longer be deployed in large formations. Of the three great military arms of the services, namely the infantry, the cavalry and the artillery, only two of them were to remain as essential services from now on. Furthermore, it was recognized that artillery with explosive projectiles was the most murderous means available on the battlefield.

This was the signal for the development of artillery as a priority. By every means, research was initiated for improvement in the power, mobility, range and rate of fire, in that order.

For a clear understanding of the information that has been compiled, it must be realized that while the cavalry was being disbanded, the infantry and the artillery sought to divide its functions between them, without always having the power to fulfil them.

For over four centuries, cavalry had been employed to perform the role of both shock-troops and reconnaissance. As shock-troops, they were used as a battering ram to break down opposition, both physically and morally, and, in doing so, to prepare for the assault by foot-soldiers. The scouting activity was used to probe the enemy flanks which were to be turned. Of these functions, the artillery was to take on part of the shock tactics by making the preparatory hammer blows leaving the infantry, on its own, to inflict the physical shock; the cavalry had, in the past, the advantage of doing both at once. It is immediately apparent that unified shock tactics resulting from the welding of artillery fire-power and infantry assault was not so simple to achieve.

The problem of scouting was critical. The cavalry continued to fulfil its periodical reconnaissance function. But, from the time of the siege of Paris it had been established that the increased range of artillery weapons enabled the enemy to be held by firepower alone over a certain distance, without the need for occupation by the infantry.

This was the situation with artillery at the beginning of this period. It was only after the heightened crises of World War I that it was recognized that fire-power alone could not replace mobility. This fact having been established, cavalry was "re-invented" by the creation of mechanical assault vehicles. This allowed the artillery to regain a better balanced function as a battle-field unit. From this time onwards, artillery was seen to gradually exceed the limits of its original concept.

So far, only a slight picture of combat conditions up to 1872 has been sketched in. An arms race invariably springs from the legitimate needs of the headquarters staff to have those arms whose necessity is envisaged by current trends of military thought. In their own language, the following technical details eloquently describe this situation. They are certainly intriguing when it is borne in mind that this was one of the first arms races of this kind to be undertaken in the modern world on the road to industrialization.

The German army had its greatest difficulties with firmly entrenched troops, as at the Cemetery of St Privat (above). To dislodge the enemy, they needed easily manoeuvrable heavy guns. The Prussians learned from this, and the development of such guns was made a priority task.

The simplicity of a battery of '75's was deceptive—it was after long and hesitant research that this method was adopted. In those days, passionate argument raged over the artillery. In default of actual combat experience, speculation was rife as to the respective merits of light QF guns as opposed to a heavy gun system.

The effect of a 240 mm shell, fired on Krupp's private proving grounds. This picture appeared in the firm's catalogue for 1890.

The forges at Krupps had a very Wagnerian aspect. Germany, then heavily industrialised, was inclined to favour ultra heavy weapons as a result.

In France

Armament manufacturers were requested to increase the range of existing materials as quickly as possible and at the least cost. The means at their disposal were far smaller than those of the large German and English manufacturers. De Reffyetire started off with the existing 75 mm and 85 mm bronze barrels, making grooves in them, similar in concept to modern rifling, and thus increased their range to 1,000 yds and 6,350 yds respectively. This was a remarkable performance for weapons made in this metal.

In order to achieve this within the planned economies imposed by the necessity for payment of the huge war debt of five billion in gold demanded by the Prussians, even the weapons made by the Vallière method were grooved in this way. The problem of sealing or blocking was resolved by a "naval" breech, opened and closed by a screw with broken threads. By forcing its terminal cone inwards, a seal was made such as could not be achieved so easily by other wedged methods. The reinforcing rings in lead on the shell were probably derived from the Whitworth method. This British system stemmed from the development of a projectile whose entire outer shell was made in this malleable metal which fitted closely into the grooves of the rifling.

In order to achieve long range, these weapons were elevated to an inclination of about 30°. As the ordinary wooden gun-carriages could not stand up to the recoil of a barrel being fired at such an angle, they had to be replaced with stronger, metal gun-carriages, whose construction could not be carried out by traditional craftsmanship methods. Recourse to industrial processes was necessary to construct them and these, also, had to be developed.

But the ingenuity employed to make use of the bronze artillery pieces did not succeed in misleading the War Minister. This was a temporary expedient of which everyone was aware and apprehension was increased by the appearance in Prussia of steel weapons having an appreciably higher performance.

In Prussia

In 1873, the Prussian High Command had just adopted the Krupp cast iron 78 mm calibre piece for horse-drawn batteries and the 88 mm calibre, called the "90", for its field artillery.

Considerable research had been devoted to the development of fairly heavy projectiles, keeping in mind the success of the explosive nose cones used in 1870. These consisted of two layers of metal in which rupture lines increased fragmentation.

Pieces of ordnance firing such heavy projectiles had a considerable recoil action, and attempts were made to limit this with a wheel brake, which also decreased the time required to re-lay the gun. However, although this contributed to the solution of the recoil problem, wheel brakes and spade trails alone could not entirely overcome the immense forces released on firing.

The breech opened and closed transversally and included a copper sealing ring, placed by hand into a housing filled with grease. This was a weak point in the system as irregular wear meant frequent changing. In this respect, the naval screw-closing taper breech was much better.

In Great Britain

The two companies of Armstrong and Whitworth were in competition as a result of their individual technical developments. On the Continent, any revelation of their new inventions and developments was closely followed. Some notable British developments were the screw-closing breech, and the polyvalent "fragmentation projectile", the same basic shell able to be used, according to the charge, either as a shell filled for fragmentation, a time shell or a percussion-time shell.

The British cannon of this period were, for the most part, forged by the "ribbon" process. This consisted of winding iron rods on to a cylinder then squeezing the tube so obtained into a second sleeve. The accuracy of these tubes complied with the specifications laid down by the Navy and was superior to that of the pieces of ordnance used by the land forces.

Krupp attentively followed these developments. But as the cast iron barrels were liable to burst this weakness was overcome by limiting the charge and slightly decreasing the muzzle velocity of the projectile, thus restricting the range.

A German 105 mm gun firing. The quality of the barrels was such that if the steel was lightly tapped with a piece of metal, it gave out a clear note like a wine glass. Its advantage was the narrow dispersion of its shots —150 ft at 5 miles.

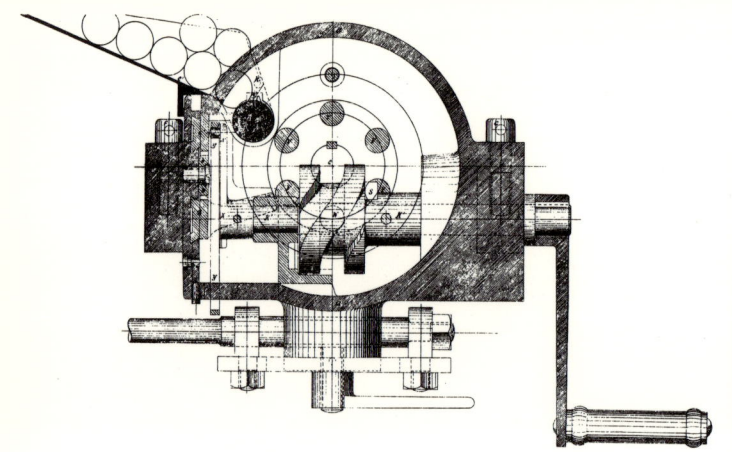

Detail (top, left) of the Hotchkiss-Broadvell revolver-cannon. In a given time it could fire more rounds at a target than a normal heavy gun, but due to weakness at the breech, it lacked penetrating power.

Above: a screw-breech mechanism. The actual discharge was effected by a groove in the breech-block. Thus the problem of a gas-tight fit was solved after a fashion, but at the expense of rapidity of fire.

◁ *57 mm QF gun of 1889.*   *Rifled 285.5 mm gun of 1873.*
▽

The Americans

One of the two largest manufacturers was a firm called Gatling who had built a weapon having eight barrels mounted in a ring. By the rotation of these barrels, reloading and firing were carried out simultaneously. The major advantage of this piece of ordnance was that re-laying was unnecessary each time as the small calibre of its barrels, a maximum of 25.5 mm, resulted in the recoil being absorbed by its stand. The Gatling method was, moreover, taken up again ninety years later by the manufacturers of the Vulcan gun, which had a rate of fire of 6,000 rounds per minute.

The other big manufacturer was headed by a certain Hotchkiss, who lived in France. He had built a type of rotating cannon with a calibre of 37 mm, able to fire eighty rounds per minute and with a range of two and a half miles. This cannon discharged a greater quantity of pig iron and fragments of shell than any other existing field piece. Many artillerymen were attracted to this piece. Its principle advantage, shared with the Gatling gun, was based on the fact that no re-laying was necessary during firing. It was not widely adopted as its trajectory was not of a low enough level and the shell not very effective against buildings, where enemy troops might take shelter.

About 1875, at the end of this first phase, France was very anxious not to be outclassed and adopted a piece of ordnance intended to be superior to the Prussian 90 mm. Commander Lahitolle's 95 mm was an iron gun, with "frets" along the greater part of its barrel length, beginning at the rear end. These frets were "... steel rings with an inside diameter less than the outside diameter of the barrel when cold..." These rings or frets were mounted when hot and, by shrinking as they cooled, were tightly locked on to the barrel. This "fretting" method was adopted by Lieutenant-Colonel Treuille de Beaulieu from 1858 onwards. Each of the nineteen army corps received two batteries of these pieces.

In order to construct a practical QF gun, it was necessary to resolve the problem of recoil. In 1895, German constructors felt they had found the answer. And then—two years later—Konrad Haussner registered a patent for a brake...

... which had certainly not escaped the notice of Krupp's engineers. Why then did they ignore it? In order to sell the "first generation" QF guns? Quite likely. However, it is hard to find justification for an artillery park full of differently constructed guns.

THE SECOND LAP OF THE RACE

Throughout this evolutionary process, it may appear erroneous to distinguish between each heat but, nevertheless, these were clearly defined. Military history has always been punctuated by the regrouping of troops following the continual change of ideas in all those situations where the requirements are such that the high command is forced to bring about a new order of troop concentrations or rethinking of basic military philosophies.

At this juncture, the technical progress had moved along on the fringe of military doctrine more or less independently. It became necessary, therefore, to re-examine military ideas on the basis of the new relation existing between the density of fire and the possibilities of attack.

Expressed in the power of explosive force, the highest performances were those of the Hotchkiss guns, a battery of six being able to fire, theoretically, 4,800 rounds per minute on to the vanguard of a battalion launching an attack, thereby halting all movement, and thus compelling the commander to change his overall tactical theories. But the Hotchkiss gun was not adopted. The field pieces in service were far from being able to provide such an avalanche of explosive power. This could only be achieved when recoil was eliminated and the power of the fired projectiles could be increased. This remained a project for the future. In the meantime, it was considered sufficient to improve those arms that existed, to extend the attack formations, to replenish stocks of the ordnance in service and to economize on the luxury of a total re-design whose necessity was, nevertheless, felt to be imminent.

In this context, the French army adopted the Bange method (see p. 149) with the whole range of its calibres. This was the last instance in the history of artillery of an "overall method" of this kind. Mountain artillery was provided with a bore of 80 mm. Mounted artillery, its gunners seated on the limbers or ammunition wagons, was issued with a piece of ordnance of 90 mm, from 1877. Short and long siege and garrison guns of 120 mm and 155 mm were also provided and, finally, mortars of 220 mm and 270 mm.

While the French were building up their field artillery *per se*, the Germans were closely studying

This German gun found a number of supporters for its "long recoil" system, but it did not mark any significant step forward. The potential buyers were well aware that the systems being tried out had not yet reached the most effective solution.

the effects of artillery on fortifications and the architecture of fortified emplacements. They had not forgotten the losses suffered from all the fiercely defended locales in 1870. Neither had they forgotten Strasburg, taken at the third breaching of the walls by excellent siege pieces, nor Belfort which had defied their efforts. Juliers, the ramparts of which had to be demolished, was submitted to a wide variety of assault tactics. The great strides made by their national industries enabled them to invest considerably in the creation and development of heavy field artillery. This fact merits consideration for it explains the origin of developments which were to have dramatic and far-reaching results.

The Russians were not out of the race. The effect of their bitter defeat suffered at Sebastopol was analysed by Todtleben with a frankness rarely equalled in works of this kind. Over the whole field of military literature it is, indeed, difficult to find anything like these four volumes, which incorporate large plates illustrating the whole development and the formation and emplacement of the guns. The commanders who suffered defeat at Crimea studied these tomes and, particularly, the passages concerned with the effect of cannon fire, with an almost clinical intensity. One of the results deriving from these post-mortems was the introduction of the "107", a very heavy piece for field artillery.

As for the English, they had the biggest possible calibres on their drawing boards, such as the 80-ton 406 mm cannon, and were preparing the manufacture of even heavier guns of one hundred tons which had been ordered by the King of Italy from the British firm of Armstrong.

During this second phase, France made every effort to gain and hold the lead in field artillery for herself. With Parliament accepting the need for rearmament, a large-scale and clearly defined research programme was initiated, concentrating the studies of the technical schools. Finally, owing to the spur applied by a military command open to new ideas, decisive steps were taken. In the year 1884, an engineer called Vieille invented gelatinized gun cotton which became better known as smokeless powder. Although this invention has been somewhat forgotten since, it wrought a considerable change to the face of the

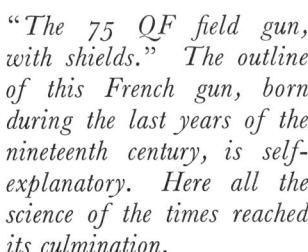

"The 75 QF field gun, with shields." The outline of this French gun, born during the last years of the nineteenth century, is self-explanatory. Here all the science of the times reached its culmination.

161

The German 150 mm heavy howitzer was a success from the first. It did not weigh more than two tons, and fired a projectile of 92 lbs over a range of 5.5 miles. The searching trajectory of this gun left few areas of dead ground.

Modern QF guns were frequently found beside a considerable collection of siege guns in the arsenals of Europe, remnants of outmoded types, like this 150 mm howitzer. It is a mistake to assume that new guns forced belligerents to adopt a war of movement.

A Russian siege gun of 1904. The Russo-Japanese war, and the Balkan wars served as proving grounds for heavy artillery. Guns that, properly speaking, were usually classified as siege-guns could be found in the role of field-pieces.

battle-field. It should be remembered that each volley from the infantry, using the old powder, created a thick smoke-screen which often allowed the cavalry troops to hurl themselves on to the infantry before they had time to reload their arms. As for cannon fire, this was comparable to a veritable bon-fire generating smoke and confusion. It was also used to camouflage tactics. The introduction of smokeless powder was not without disadvantages. Until the appearance of smoke shells, the artillery could no longer create those very practical smoke-screens behind which troop movements could be concealed from the sight of the enemy.

As usual, in peace time there is scarcely any thought of camouflage and it is happily imagined that in time of war the terrain itself could provide the defender with sufficient cover. Thus, in 1890, it became the practice to fire from natural palissades over the troops.

Just as in other periods of history, the end of the nineteenth century also heard arguments between the old and the new brigade. The older partisans considered that firing from concealed positions was quite contrary to the upholding of military ethics. Their opponents, on the other hand, claimed that it would provide an extraordinarily wide range of pos-sibilities. In defence, the Old Brigade quoted tac-tics employed in recent battles between Russian and Turk in 1878, during which the Tsar's armies, faith-ful to their military tradition, had insisted that their artillery had the physical support of the infantry, and their open participation in the assault. To listen to them, the success of the Russian armies was sufficient proof of the uselessness of concealed fire. On look-ing back, it is quite easy to understand how the legacy of a certain cavalier spirit carried so much weight in artillery circles of that period. Nevertheless, the stand taken by the old brigade did not prevent the out-spoken expression of more modern ideas, nor entail its exponents being ostracized. This period, during which all ideas were given free rein without being stifled by dogmatism, is one of the most interesting to study, as is any period immediately prior to the re-birth of a weapon. And now the technical experts were hard at work on the problems of overcoming recoil.

THE THIRD LAP

The outstanding achievement of the period from 1894 was unquestionably the taming of recoil. The adversaries in the lists, the French and Germans, were closely matched in their endeavours, their research being conducted along parallel lines. Sooner or later, they would arrive at similar solutions.

As always, the intelligence services played their role. Today, little mention would be made of their contribution had not the Dreyfus affair given this technical problem a singularly dramatic aspect with which it is irrevocably linked in retrospect.

The idea of making a moving tube slide on a fixed cradle and of absorbing its recoil by a hydraulic brake is neither new nor truly secret. Its application, however, encountered innumerable difficulties. In fact, this brake is none other than a kind of large shock-absorber, the same principle which has been used so widely ever since in motor vehicles. It consists of a piston, connected to the tube which is provided with an opening, sliding in a cylinder filled with a liquid. Theoretically, the function seems quite simple; in practice, this is not the case. During tests

on the first type, the brake broke down at certain elevations as the weight of the tube increased the recoil. A means of gradually altering the size of the opening in the piston had to be found. Moreover, at certain temperatures, the composition of the liquid had a tendency to emulsify and the decrease in the level of viscosity completely upset the calculations made for the opening in the piston. There were countless difficulties to be resolved but engineers got to the bottom of their problems quite quickly.

For instance, in 1891, a German named Haussner filed a patent in the German patent office, covering a long recoiling brake intended for field pieces. It can still be seen.

The President of the *Comité Technique de l'Artillerie*, General de la Hitte, immediately recognized the importance of this invention. He ordered the *Fonderies de Bourges* and the *Ateliers de Puteaux* to study its applications without delay.

At Bourges, however, this project was shelved and returned without examination after two months. The appearance of the Haussner patent was considered highly inopportune by the director of the foundry, as his organization had decided on the construction of a field piece with a short recoil.

At Puteaux, on the other hand, the German invention was turned to full account. Owing to the energetic drive initiated by Commander Duport, in only one year the workshops had produced a prototype of this "75" field piece which was to play such a major role in the future. No time was lost over the work and the secret of its construction was maintained by allocating the manufacture of different parts, whose precise and ultimate function was difficult to guess, to numerous sub-contractors.

This piece of industrial deception was successful and the various other stratagems employed only strengthened the Germans' conviction that a short recoil was much more valuable than the long recoil. The patented Haussner long recoil nevertheless did not escape them. The trial of Captain Dreyfus, convicted three years after the filing of the patent, did not alter the fact that its inventor had not succeeded in capturing the interest of arms producers in his native country.

When the Germans finally realized the advantages of a long recoil, it was too late to include it in their basic weapons system and they were obliged to reconvert their field guns with all the disadvantages this entailed. If, as many people expected, war had broken out in the first few years of the century, the Germans would undoubtedly have suffered a serious setback from this oversight. As it turned out, by 1914, they had had ample time to remedy their mistake and their subsequent weakness was due, not to their guns, but to their ammunition which did not match their guns in performance.

On the other hand, there is little doubt that the Dreyfus affair was skilfully utilized by certain French politicians as an "open sore" to arouse public opinion. 1898 saw the sensational reopening of the trial of the unfortunate Alsatian captain, and the unobtrusive introduction of the new arm into service. The enormous sum of three hundred million gold francs had been spent on perfecting this weapon. Public opinion in a country with a strong traditional attachment to the land would never

willingly accept expenditure in investments of this kind. It was thus hoped that the Dreyfus affair would serve to divert the public's attention from this outlay. But the proof of this surmise has never been established, as is so often the case in State affairs.

One might be tempted to overlook this strange business, whose romantic flavour does not seem to have contributed very much to the history of artillery, had it not also revealed two new and significant facts concerning one of the first very large series of modern cannon.

The first of these facts was the now obvious necessity for governments henceforth to coordinate defence plans and budgets more closely with research programmes and, especially, with information channels where the weaknesses had proved greatest. It was all because of a gun that public opinion, prematurely aroused to an almost warlike pitch, hardened and, worse still, became dissident.

The second important fact is the extraordinary simplicity of a weapon produced from studies and research which were, in fact, highly complex. The proof of this can be found even today, in the fact that we accept the appearance of this piece of ordnance as obvious.

The advantages of this gun were as follows:

1. Simplicity of operation enabled gunners to be trained to a professional standard in a relatively short time.

2. Simplicity of firing enabled the officer commanding to adapt the trajectories to any kind of terrain whatsoever. There was even a standard charge unit by which degrees of fusing the shell were made identical. This enabled accurate and quite rapid fire to be directed on to enemy infantry troops, who could no longer escape out of its range.

3. The ammunition was better to handle. A well-trained gun crew could easily fire twenty-five to thirty shells a minute.

Elementary as these advantages may appear, they actually represent the outcome of highly involved calculations whose supreme refinement lies in the simplicity of their application.

In short, this is one of those rare engines of war that has never been given the time and resources to fully develop. As far as the troops were concerned and, unfortunately, the high command as well, it was a passing craze. The "75" was the favourite of the day and, just as forty years before with the Chassepot, the "75" was the maid of all work.

Admittedly, there was no lack of voices raised in criticism of the "75's" limitations. Curiously enough, however, it seemed that the adoption of this weapon had, at the same time, engendered a myth as to its invincibility.

It was at this point that the proceedings became even more dramatic than had been the case over the Dreyfus affair. Quite simply, artillery had brought about a revolution of which the infantry had seemed totally unaware. From now on, the artillery could saturate the fringe of the attacking forces with explosive fire and pin them to the ground because of rapid fire from its batteries. The British, however, had learned a great deal from the Boer War, particularly the uses of cover for artillery, indirect fire, and the advantages of open order for the infantry.

Faced with the increased fire-power from enemy artillery, supported by machine guns which had also just been perfected, the infantry had to totally reorganize and re-think its methods of attack and engagement. This applied as much to the French as to the German infantry, the introduction of the "77" into the Kaiser's armies being more or less equivalent to that of the "75" in France. Strangely enough, the infantry scarcely altered its formations in the field. They still attacked in closed order lines. The number of wounded from the gunners' shellfire provided eloquent testimony to its efficiency. But that drastic school, war, ended all discussion.

In his writings, Foch laid down the basic principles of warfare. "The percentages obtained in target practice, the effectiveness of artillery fire as seen on the ranges, makes attack appear fundamentally impossible. It is therefore necessary to avoid attacking and to invite attack, to return to a war of position and skilful manoeuvre, to circumvent or avoid the enemy and to starve him out. Each improvement in arms confirms the need for a return to defensive warfare. The same questions, studied from the opposite point of view, with a history book in one's hands, suggest the opposite reply". This opposite reply was the offensive at any price, whatever it cost to preserve the legend. The origins of this strange and stubborn short-sightedness would alone merit a whole book on the subject.

In 1919, when the recent tragedy had filled Europe with rubble and tears, General Gascouin, a great artilleryman, explained the matter in these terms: "Although, fifteen or twenty years before the war, the Staff College used to perform a constructive role as a military critic, it has become, especially since the creation of the Army Council, a simple information service for the official point of view... Sometimes, in such teaching circles, trends of thought unrelated to reality take root... periods of mental contagion in which, by definition, *opposition* and *controversy* are banned".

The introduction of the French "75" gun has been discussed at length because it is particularly representative of the evolution of artillery in all armies at this period. A brief general look should now be taken at some other weapons before they enter the field.

In Germany, the "77" gun, similar to the French "75", was adopted by the army for its field artillery. These pieces were formed into batteries of six guns while the French, for reasons of greater flexibility, formed them into units of four guns.

The German artillery also had a light "105" howitzer for raking fire, and was completed by a park of heavy pieces which is discussed in more detail in the study of the conflict itself.

Moreover, on his own initiative, Krupp had developed heavy mortars of 420 mm, intended to smash through the concrete of the strongest fortifications existing in Europe at that time.

Before the First World War when the weaponry and armaments were very soon going to face each other, their methods of engagement were put to the test in two other theatres of war.

The Russo-Japanese war, which, with the Boer War demonstrated the necessity for concealed fire,

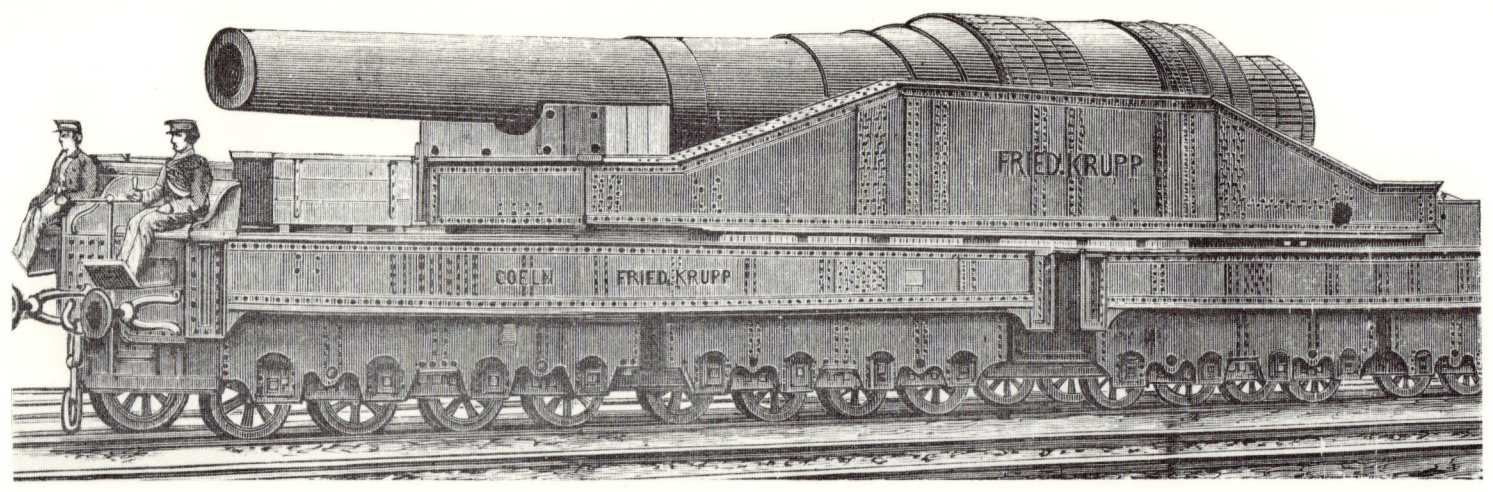

In the middle of 1885, the Krupp works successfully made 4 15.7 in guns. The barrel alone of these gigantic guns weighed 121 tons, and a special waggon had to be built to carry it. Its projectile weighed 202.8 lbs and was 4 ft 2 ins long.

which was going to proliferate more and more in the European armies. The machine gun also made its true operational appearance and its importance was not overlooked by informed observers.

But the fate of intelligence officers will always be similar to that of Cassandra. The more far-sighted they are, the less attention is paid to them. In fact, European armies were more interested in the rapid-fire cannon and in maintaining the offensive spirit of the infantry than in the machine-gun. Nevertheless, the Austrians and the Swiss did devote more attention to this weapon and, particularly in Switzerland, the future General Wille formed veritable batteries of limbered machine-guns well in advance of their time. The Balkan wars also enabled the Austrians to test out their heavy field artillery.

It is well worth while to take more time than is now available to make a deeper study of the evolution of artillery as used in fortifications. Suffice it to recall the some 4,000 towers and casemates of all types which defended strategic points and frontiers of the European nations at that time. The park of "garrison" pieces of ordnance was even greater, being three or four times higher in number than the towers and casemates, as the old cannon were not destroyed without reluctance or good reason. Their number is very high if compared with the much more modest lists of field artillery units. This helps to explain why, in later years, belligerents brought

back into use some of the ordnance with which fortresses such as Verdun had been endowed.

Also drawing on the forgotten past, the original invention of a Swiss, the famous Colonel Julius Meyer, is worthy of note. He was preoccupied with the large-scale development of heavy artillery and sought to provide a counter-weapon whose cost was within the range of his country's finances. Inventors are always somewhat visionary. From 1891, Meyer foresaw the possibility of a long siege war. Therefore, he set himself the task of producing a piece of fortress artillery which could maintain its effectiveness despite the heaviest bombardment and actually designed a method of small, mobile armoured casemates. These were in the shape of steel bells, hardly six feet in circumference, which would be sheltered from the fire of the enemy guns in tunnels, made in the rocks, during the initial stages of the siege. When the artillery ceased fire in order to allow the infantry to mount an assault, these steel bells, rolling on Decauville rails, made their appearance in several seconds and swept the slopes with their rapid-firing 54 mm gun. This curious antecedent of an armoured assault vehicle not only met with approval but it was a veritable weapon system, perfectly suited to the peculiarities of the Swiss terrain and, more particularly, to the Swiss budget. The bastioned frontal walls of many Swiss fortifications were equipped with this weapon.

THE RETURN OF SIEGE PIECES (1914-1918)

The threshold of the great world conflict was just ahead and all the belligerents were to throw themselves into the war with the conviction of winning it.

This conviction was mainly based on the effectiveness of their artillery. The British concentrated, first and foremost, on artillery for their ships of the line, which numerically outclassed all the navies of the world. For example, the *Dreadnought*, built in the space of only a few mouths in 1906, is a typical illustration of the Anglo-Saxon style. The aim had been to mount a truly vast range of guns on a vessel especially built for this purpose. But as they had not yet resolved the problem of protection against torpedoes, this system of massive fire-power was not totally "integrated". The French relied on the strength of their field artillery which they had made the most manoeuvrable in the whole of Europe.

The Germans based their confidence in the shattering numerical superiority of their medium and heavy artillery of 105 mm and 150 mm. Totalling 3,500 pieces, 2,000 were of 150 mm. Against these, the French had only 300. In addition, the Germans were equipped with 420 mm mortars against which no fortification could hold out.

The future belligerents were not totally in the dark about reciprocal hypotheses and theories. But they tended to rely too much on their enemies having the same reactions as themselves—the only "enemies" encountered so far being their comrades who adopted this role in tactical exercises.

The French expected to see the Germans seeking an energetic rather than a tactical combat. They anticipated a sudden offensive without ulterior motive, nor big reserves, and where "the violence of the act and contempt for the enemy always compensate for errors".

Primed about the onrush of the German troops, it was planned to direct them into a veritable trap where the artillery could annihilate them.

For their part, the Germans were confident in the force of their formidable battering ram. They were aware that their superiority of fire-power boosted the morale of their troops who had confidence in a high command with large resources at its disposal. They accordingly prepared to take the

24 March 1916—a battery parades in a review near Bar-le-Duc. It is a perfect representation of the French nation at this time. The drivers and their horses came from the country, the magnificent machinery from the town, while the elite bestrode their chargers. The commander, quite likely, came from the Polytechnique.

After the victory of the Marne, which was aided by the devastating effect of the gun, a myth grew round "Notre glorieux 75". But myths are like tools—they are used, for a time, and then they have to be replaced.

These Italian gunners manhandling a gun up to a wooded ridge show what kind of obstacle could be surmounted by a well-trained crew.

Munition columns had to move up in tune with the demands of the guns they served. Mud was the great enemy of the convoys, and until the coming of tracked and armoured vehicles, provided much hardship and heartbreak for the troops.

The specialised needs called into existence by trench warfare called for a diversification of functions for guns that were not originally envisaged in their classifications. In this picture, a 90 mm mountain gun has been used as a fortress piece in an improvised fort.

Russian gunners in 1916. The Pontilov 76 mm was an advanced gun for its time, but the Tsarist armies had too few. The artillery superiority of the Central Powers allowed them to tear holes in the Russian front.

offensive, convinced that by maintaining the initiative, they could reduce the French to a state of paralysis. If, at that time, anyone had talked to them about traps, they would doubtless have replied that, under the blows of a good battering ram, no trap could hold them.

Nevertheless, at the last moment, the need to carry out several changes was felt in both armies.

Changes to the Schlieffen plan in Germany, whose main aim was thereby undermined, were perhaps due to the huge increase of 50 % in the French artillery which was decided on in 1909. This fact is difficult to prove but is a possible hypothesis.

In France, General de Lamothe was recalled to service. This seems strange but typically practical when it is remembered that this officer had fallen into disfavour in 1910 because he had insisted too strongly on a substantial development of heavy artillery, but his ideas were now accepted.

Thus, at last, the question of fire-power was being tackled. But too little time now remained to solve this problem and the high command, overloaded with various projects, overlooked a very simple solution which could have re-established the balance of arms and munitions and, perhaps, have prevented the war. A very simple and ingenious gun-carriage had been designed by Major Mourcet at Bourges. It would have been quite simple to use it for

A German gun in action in Champagne, 1918. A growing awareness of the uses of mobility in counter-battery and defence work led the Germans to stop building elaborate emplacements.

This French 75 is mounted on an improvised pivot mounting for antiaircraft work. The picture was taken in April 1915, proof that the aeroplane was making itself felt.

mounting a thousand Bange 120 pieces, for which there existed a large stock of excellent ammunition.

But in such times of emergency, when the spirit of invention is violently stimulated, there is often such a profusion of techniques that the choice is difficult. Thus, instead of taking advantage of the Bange guns, it was considered better to speed up the development of a 105 gun which was still in its early stages and not yet ready for production.

Supplies of heavy artillery in the French artillery were deficient from 1913 to 1914. The equipment was only built up slowly. It was the imperative necessities of battle which led the French to equip their heavy artillery with a park of 1,500 to 2,000 pieces of 105 mm as late as the first months of 1918 and it was only in 1918, after four years of battle, that the balance of weapons, which might possibly have prevented war, was restored.

ARTILLERY IN THE FIRST YEARS OF THE WAR

The great battles of this war are too well known to need enumerating here. Our studies can more profitably concentrate on the constant and alarming despoiling in which the antagonists so ruthlessly engaged throughout this disastrous war.

The Year 1914

At the opening of hostilities, the Germans behaved exactly as predicted. Heavy 420 mm howitzers appeared on the scene. Technologically, they created considerable surprise. Their shells, weighing nearly a ton, could blow up the very solid fortifications of Liège. The battering ram was at work.

On the French side, the "75" and, particularly, those who manned it evinced astonishing flexibility. Among its admirers was an authoritative adversary, none other than the future Field Marshal Rommel, then a lieutenant and a company commander. He wrote: "... during only one engagement, the regiment has lost a quarter of its officers and a seventh of its strength... the heaviest losses were caused by the French artillery..."

Although it is simple to talk about flexibility not an engagement, the nature of this flexibility is

The lack of a heavy artillery has deeply felt by the French. The factories laboured for three years to fill the gap, but while new guns were being prepared, replaceable barrels were improvised, as with this 120 mm gun, a version of the Holzapfel D.F.

The effect of counter-battery work. This German 150 mm gun, near the Jerlan-près-Ostel farm, will not fire again. In a way, the artilleries of the time cancelled each other out, contributing to the long stalemate that reigned on the Western Front for so many years.

not necessarily apparent and requires explanation. Efficient artillery fire relies on the following basic requirements:

a) a rapid and accurate assessment of the enemy positions. Depending on knowledge of the terrain, the quality of maps and the ability of the fire commander, this stage can take from thirty seconds to several minutes.

b) ballistic calculations.

c) a list of recognized orders.

d) adjustment or ranging of fire.

All this takes a certain time. During peaceful conditions, on a fixed target, it takes a quarter of an hour for a beginner and five minutes for a good commander; a real ace can do it three minutes.

During the first weeks of the campaign in 1914, it was not so much a matter of fighting for fixed objectives as of pinning down an enemy infantry which was on the move. This situation demanded a rare capacity on the part of artillery officers for "mental arithmetic" on the spot, or a very rapid practical application of spatial geometry. The French artillerymen, many of whom had been trained in the technical schools and had considerable experience, surprised their adversaries by their virtuosity. It was they who made it possible for the orderly retreat of the armies after the stalemate of the battle of the frontiers.

But the artilleryman's weapon was not only his gun, it was also his projectile. The first months of this war were going to reveal to the Germans their

The later battles of the First World War generally opened with a gigantic artillery preparation, directed against the enemy's known battery positions. Each position and strong point was treated to a veritable rain of shells, forcing him to either go underground, or change his position Here the Germans are moving.

shortcomings in this respect and for which they would pay a heavy penalty. Their 77 mm projectile contained four times less explosive than the French 75 mm shell and proved far less destructive in use. The gunners somewhat bluntly described it as having an inadequate "yield". This "yield" relates to the weight of the explosive of the projectile in comparison to the weight of its metallic skin. To compensate for this deficiency, the Germans had to equip field batteries with six guns which made manoeuvring more difficult and seriously decreased the number of firing units, or batteries.

While the French army corps could put a total of thirty batteries into the line, consisting of twelve used as artillery units and eighteen divided between the two divisions, the German army corps could only mount eight batteries.

Developed amid great secrecy, the 420 mm howitzer enabled the Germans to smash Liège. The allied intelligence services had no idea that what was called a "short Naval Gun" would in fact be for land use. It was less successful at Verdun.

A special tractor and carriage was designed to carry the enormous 420 mm howitzer. The carriage and the barrel etc. had to be separated, as no road vehicle could cope with the great weight of the complete gun. Shells arriving on target were compared by the survivors to the noise of an express crossing a steel viaduct at speed.

The 420 mm howitzer dismantled for rail travel. From right to left: the engine, the bed, the barrel (22 tons), the cradle (15 tons), the mounting (20 tons), the base (12 tons) and the crane (7 tons).

On both sides, the calculations on the manufacture and stocks of munitions had been insufficient. In his memoirs, Admiral Tirpitz reveals that by the autumn of 1914 the German artillery had been almost totally depleted, and it was only by diverting massive stocks of armaments, intended for the German navy, that a crisis was averted.

In the first few months, a whole series of miscalculations came to light, some of which seem almost inexplicable in retrospect. To take only one example, it is difficult to understand why the Germans undertook the manufacture of munitions with such a low yield when the effects of a shell, such as measurement of the speed and number of bursts, can be methodically studied on the artillery ranges.

Although knowledge acquired to date does not provide a definite answer to this enigma, two hypotheses deserve mention.

1. It is quite possible that the analysis of experiences in the wars of 1870 and, latterly, in the Russo-Japanese operations, had a considerable influence. During these two wars, troops in the field were seen to throw themselves at positions which caused great difficulties because they were impossible to bring down. In 1870, the German time-fuses had not enabled the Prussian Guards to bring down the walls of the Saint-Privat cemetery. *In twenty minutes, these élite troops lost eight thousand men for the lack of armour-piercing shells.* It should be remembered that it was the lieutenants of the 1870 war who chose the munitions in 1914.

2. It would appear that in 1914 the German chemical industry did not have the knowledge of the processes or the installations available to manufacture synthetic saltpetre on a large scale. In peace-time, natural saltpetre was imported directly from Chile and the cost of

raw material was certainly much lower than that required for producing the synthetic variety.

The Naval blockade quickly forced the German chemical industry to find alternative means. Spurred on by vital necessity, the industry was extended and branched out into the production of gas. As a result, gas warfare came into being.

This particular aspect of artillery helps to explain the fundamental reason for the countless miscalculations of the period. As a matter of fact, in 1914,

A variation of the German 420 mm howitzer on patent, weight distributing wheels. Heavy artillery broke up the allied offensives of 1915 and 1916, and shattered the Bohemian fort of Przemysl. To slow down the German advance of 1914, the 75 mm was sufficient, but as the war progressed commanders sought to strengthen the attack of men by a rain of fire.

A well-camouflaged German gun on the edge of a wood, probably "Big Bertha", which fired 277 lbs shells over a distance of 80 miles, bombarding Paris.

A dump of 400 mm shells on the platform of Petit-Blangy station, 11 August 1916. At the time, it was estimated that it took a ton and a half of metal to kill one man, but in 1941 the figure had come down to 300 lbs.

An Austrian 305 mm howitzer. The Austrians were famous for their monster guns—but they became more and more common, on either side, as anti-personnel weapons, as the war progressed. Guns were reserved mainly for antibattery work.

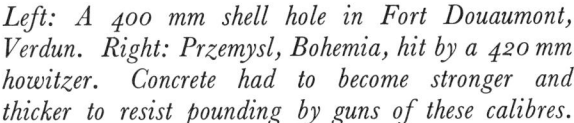

Left: A 400 mm shell hole in Fort Douaumont, Verdun. Right: Przemysl, Bohemia, hit by a 420 mm howitzer. Concrete had to become stronger and thicker to resist pounding by guns of these calibres.

the inter-dependence of industry and national defence requirements was scarcely recognized. It was not understood that an industrialized country must defend its factories with the machines they produce, and not by the sacrifice of its youth and its agricultural population. It was forgotten that a disjointed development of economic, demographic and military functions always produces a crisis.

By a process of fusion, whose significance is only now beginning to be understood, war was to restore a strange coherence where none had existed before, and teach what was hitherto unknown. Then there existed only an instinctive foreboding.

The Years of 1915 and 1916

The belligerents spent the winter feverishly striving to overcome those handicaps in equipment which had reduced to nothing some of their projects during the past year. But it was all the more difficult for them to anticipate the turn of events with any accuracy as the character of the war had almost totally changed. What would have been practical in the war of movement of September 1914 was not necessarily of use in the static war of 1915. On both sides, efforts were being made to break the stalemate and recover the mobility for which the main part of the artillery had been designed. Above all, during

this time of uncertainty, efforts were made to improve what equipment existed and had proved its worth, both as far as munitions and armaments in general were concerned. As for the cannon, there was a general drive to increase their number and thereby the density of fire per mile, rather than to modifying their type. During a war, it is indeed impossible to effect fundamental changes in the artillery with which it is being fought.

The Germans, however, were about to adopt a new line of action. Having the most powerful chemical industry in Europe, they were going to try and bring about a decisive change by a "technological breakthrough", the "big push" on friend or foe's lips.

Large stocks of gas were built up but its technical and operational use was not integrated. They started to use gas by simply opening up steel gas bottles. If gas and shells had been combined, the result might have been decisive. However, the integration of production techniques with combat use is the military proceeding that costs most effort and time. When opening their rows of gas bottles at Ypres on 22 April 1915, the German chemists realized too late that only artillery could give their lethal mists the necessary depth of penetration. To just wait for the winds of heaven to be blowing in the right direction was not enough.

Guns searched out the ground inch by inch. Here, two Germans examine the wreckage of a gun, caught out in the open by the enemy's guns. Counter battery work was one of the gunner's tasks before any attack was launched.

One of the outstanding artillery developments to take place during the First World War was the employment of guns to lay down a creeping barrage, a term used to describe the fire dropped in front of advancing troops, which was lifted from line to line on a time basis while the infantry followed under its cover. This was not a new concept, for Graham notes that at the siege of San Sebastian in 1813: "Our infantry advanced to an assault with the guns firing over their heads and lifting their fire at the moment the final charge was made; a feat which proved the skill of the gunners as well as the accuracy of their pieces." The barrage replaced the use of the cavalry in demoralising and confusing the enemy before the infantry assault. The advent of the tank, however, once again caused a reversion to older ideas, and armour replaced the guns which had, in their turn, replaced the cavalry.

The Allied High Command was also seeking a break-through, which alone would enable them to resume the offensive and to bring matters to a head by leaning on logistics. Their principal effort was directed to the manufacture of ammunition for the guns and this resulted in incredibly large stocks.

With only one common thought, the Allied artillerymen planned a systematic softening-up, during which it was estimated that if each square yard was pounded by shells this must, from the mathematical point of view, guarantee success.

As soon as the necessary amount of ammunition was available, Joffre launched a first offensive in February. The much sought break-through nearly succeeded. It seemed a near thing and it was decided that, next time, nothing would be spared in an all-out effort. This resulted in the tragedy of September 1915 in the Champagne district of France. It is less talked about than the Battle of Verdun, probably because it was worse.

Joffre chose the flat sector bordered by the Catalaunic Fields to the north and not far from Valmy as a point for his offensive thrust. This country, which had also felt the hoofs of huge troops of cavalry pounding its surface in battle, was equally suited to the heavy batteries of artillery. Joffre intended to use 2,000 pieces of ordnance over a front of 25 miles. Three railways were actually built to ensure a continual supply for this impressive array of guns.

The softening-up was to last three days. It was to be followed by infantry attacking in waves, preceded by a creeping barrage overwhelming the remaining enemy and engulfing the breach.

It was planned to submerge the Germans under an avalanche of fire and a solid mass of fighting men. Thirty years before Hiroshima, the explosive fire could already have been measured in kilotons.

As for the troop strength engaged in the battle, this was also enormous. In order to break out on these twenty-five miles of front, Joffre had brought up eleven army corps. Facing them were 2 German army corps and 600 pieces of ordnance. Logical reason suggests the French offensive could not fail.

On 22 September 1915, the offensive broke. The artillery preparation ploughed up the ground and smashed through the first German position. Three German regiments were annihilated. Now the rain decided to take an active part. The churned-up ground was transformed into one vast sea of mud. Nevertheless, after enormous efforts, the French infantry managed to reach and partly overrun the second German position on the first day.

It was then that the drama reached its climax. The French "75" was not a siege piece. And this was, in fact, a state of siege, for which very heavy cannon with curving trajectories were required. The French only had a few such pieces. As a result, their infantry had to tackle the far slope without the support of artillery fire, for the low trajectory of their "75's" was ineffectual. Here, the French troops were welcomed with howitzer fire from the 150 and mortar fire from the 120 whose curved trajectories had practically no dead angles. The shells made frightful gaps in their ranks. The weather was so thick that the airborne artillery observers could not go up in their captive balloons. It was impossible to spot the enemy batteries.

Nevertheless, the French infantry renewed their attacks with a relentless tenacity rarely equalled in history. Unknown to them, this caused the greatest concern at General Falkenhayn's headquarters. But there was no break-through. After fifteen days of fruitless attempts, the offensive petered out at the very moment that General Falkenhayn had himself decided on a withdrawal.

After fifteen days' fighting, heavy artillery had won the defensive battle. Heavy guns alone have this advantage of delivering tons of explosives on the troops in the first moments and taking them by surprise. The light gun, in spite of its rapid fire, cannot produce such a moral or physical effect, in which the first blow is decisive. Biologist René Quinton, then an Artillery Commander, wrote: "Fire that kills needs the element of surprise."

At the end of this squalid battle, a quarter of a million men were dead. Two thirds of them were French. These figures are, in no way, exaggerated. They have been taken from the reports of those armies which suffered the losses and which had nothing to gain by increasing them. The siege piece had made its comeback. But the very substance of the old Europe was heavily stricken by this bloodshed.

A British pill-box destroyed by artillery. "The one you hear won't get you" was a well known saying of the time. Not only would it not kill, but also the enemy shells sometimes helped the defenders by creating obstacles difficult to traverse.

The war of attrition continued. General Falkenhayn knew that the French offensive of September 1915 had very nearly succeeded. Convinced, in his turn, that by employing adequate resources, such an operation should result in victory and aware, moreover, of the superiority of his heavy artillery, he planned the Verdun offensive.

The ground would not merely be ploughed up but literally excavated, one heavy shell blasting every four square inches in order to blow the ground out from under the very feet of the French.

It was on these strategic and technical calculations, in which artillery took the leading role, that the fate of the German Empire was going to depend. But irrational, explosive tactics such as these cannot be determined by calculations alone. The 420 mm guns which blew up Liège and Przemysl battered the towers of Douaumont and Moulainville without success. At the height of the earth-quaking uproar, the impregnability of the concrete and rock allowed one of the defenders to cool-headedly note that a shell from a 420, landing on the fort, made a noise comparable to an express train crossing an iron viaduct.

At the climax of the crisis, siege artillery thus discovered its limitations. There were arches that proved unshakeable, towers that stood up to tremendous battering, landslips that facilitated the defenders' task and whence emerged totally unexpected fire from isolated machine-guns which determined the outcome of the combat.

It is now estimated that one and a half tons of metal must be projected to kill one man. Quinton wrote: "One of the woes of static warfare is that those who command cannot see and those who see do not command".

The Years 1917 and 1918

The year 1917 is important in that it resulted in arms manufacturers diversifying the use of their steel. It was found more profitable to devote some of it to armouring tanks rather than investing it all in munitions.

At the same time, cavalry and its mounted batteries were reinvented. But this was done on too small a scale to show conclusive results. Knowing how many new weapons to produce for a campaign has always been one of the most difficult factors in the art of war.

As the tank was still too slow, the artilleryman ended the war with the conviction that he had the upper hand over these half-blind iron mastodons. As for the artillery itself, the battles in 1918 were a repetition, perhaps with technical improvements, of those in 1915.

This war ended with the same pieces of ordnance being in service as those that had existed at its outbreak. The French had made good their deficiencies in heavy guns and the Germans their field equipment. Ranges had been increased, gas-filled shells had been produced and guns had been mounted on tracked waggons.

But although the guns were the same, artillery itself had undergone a radical mutation. Methods of sound ranging, telemetry, map-making, aerial observation and transmission of firing information had provided the High Command with a veritable control panel. In other words, although artillery was now considered as less consequential than armoured vehicles, it had acquired even greater importance by the assistance it could henceforth provide to field intelligence services. In this respect, the navy, whose ships were equipped with the most advanced radio, telephone and transmission systems, had helped a great deal. It should be recalled that even in 1914, the gunners on land still mistrusted the telephone, considering it unreliable and even detrimental to the maintenance of the old military virtues. In order to enable a semblance of indirect fire, field batteries in 1914 were provided with two telephones and 500 yds of cable. By 1918, however, without the thousands of yards of wire laid by the gunners, the High Command would have been unable to operate. Artillery had recovered its vocation of providing an overall view without which its fire is of no avail. But whether its commanders were aware of this unique advantage is another matter.

* * *

Shell fire transformed fields, villages and forests alike into a desert that closely resembled some weird lunar landscape "where no birds sang".

DIVERSIFICATION OF ARTILLERY BETWEEN 1918-1945

At the end of the war, most of the arms had been widely diversified. The use of the cannon had spread everywhere. There were infantry mortars or *Minenwerfer*, anti-aircraft guns and anti-tank guns, all derived from the basic artillery. The air force, as yet unable to utilize the army's guns had, however, asked for shells to use on their first bombing raids.

The profusion of resources did not help to clarify ideas on their use. This was unfortunate as the slashing of military budgets obliged those in command to make a difficult choice.

Fifty years later, it is easy to distinguish between the essential and the non-essential. It is always simple to see, in retrospect, what should have been done. Our concern is to reveal the doubts of a troubled period in which choice was often, perforce, replaced by compromise.

The most important legacy from the First World War is, unquestionably, the tank. Armoured assault vehicles gave mobile warfare a new lease of life and restored cavalry to the battlefield.

Even so, military experts and general staff headquarters in 1920 could only dimly foresee the inevitable, long-term repercussions on the artillery of the introduction of this new, young armoured corps. Their appreciation of the advantages of artillery over tanks was based on these points:

a) *In many instances, artillery seemed to have proved its superiority over the tank.*

Judging from recent experience, gunners very often had the advantage over these lumbering, halfblind steel giants, easily picked off with an AA gun which was easy to aim.

b) *Tanks cost very much more than guns.*

During the war, the price of a tank was some £ 400 at the gold value, per ton. After the war, the price shot up. Modernizing the guns was much less costly and seemed at the time to be better value for money.

c) *Tanks were far from being fully developed.*

Even a brief look at the materials in use at that time is sufficient evidence.

d) *Memories of battles fought are the most persistent.*

It is difficult to dismiss the idea of static warfare and almost impossible to conceive a modern type of mobile combat in its place.

e) *The necessity of making use of an existing, huge artillery park obstructed progress.*

All these factors blinded military experts of this period to the undoubted advantages of using tanks:

— *Armoured vehicles would once again enable the use of the three basic arms,* i.e. infantry, cavalry and artillery. The air force would fulfil the former function of the hussars, as well as some of the previous functions of the long-range artillery.

— *Both artillery and infantry would be relieved* of many tasks with which they had been encumbered since the disbanding of the cavalry, particularly the transport of troops and equipment.

— *Artillery could again fulfil a specific and unique function.* As well as transporting tons of munitions by rail, and on its own carriages, artillery was also called on to assume partial responsibility for troop movements as well.

But what was this specific and unique function? It was to provide headquarters with general intelligence, for which it was admirably suited. To go into action rapidly, whether by night or day, and regardless of the weather, at those critical points disclosed by its reconnaissance services.

For artillery was not only an arm of the military services but also a mental discipline. It obliged the commanders, from their earliest years, to engage and effectively control guns and gunners that were out of sight. The artilleryman must, therefore, be

At the end of the First World War, Lt Colonel Rimailho and other French gunners turned out ideas for self-propelled guns. The St. Chamond works manufactured prototypes of a 280 mm, 120 mm and 194 mm gun.

not only a good commander but also a skilful tactician to be a good soldier.

This highly specialized discipline trains the artilleryman to make a rapid overall survey in order to position his fire at strategic points where, like a lever under a load, it will have the most effect.

In order to adapt its role to post-war conditions, artillery had to be well-informed and equipped with resources that were both flexible and powerful. Its requirements were the following:

— An artillery reconnaissance service having at its disposal observation aircraft with efficient fighter protection.
— Reliable radio telecommunications.

— Guns with a minimum calibre of 105 mm.
— Motor tractors and gun-carriages with tracks.

This equipment had been clearly specified by first-rate gunners who had won fame in the war. Prototypes of artillery with tracks had already been produced. But as complete revision of the entire system would cost too much, improvements were made in dribs and drabs.

Anti-tank guns equipped for rapid aiming and firing were produced. Muzzle-brakes were introduced, enabling recoil to be limited and range to be increased. For the sake of simplicity and, particularly, economy, horse-drawn artillery was preserved with the idea that it might still be useful,

Left: St. Chamond 75 mm SP gun, with both track and wheel systems.   *Right: The same gun with wheels retracted.*

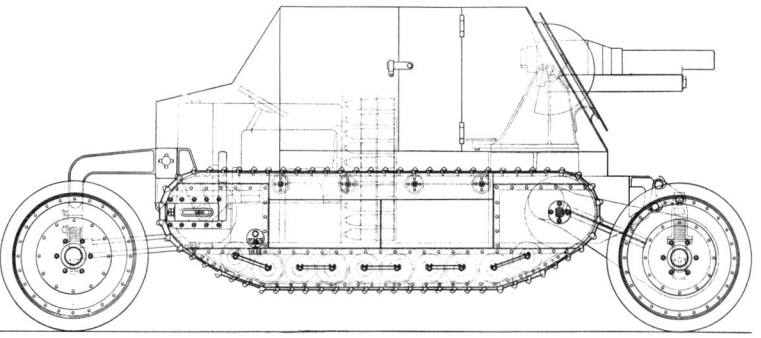
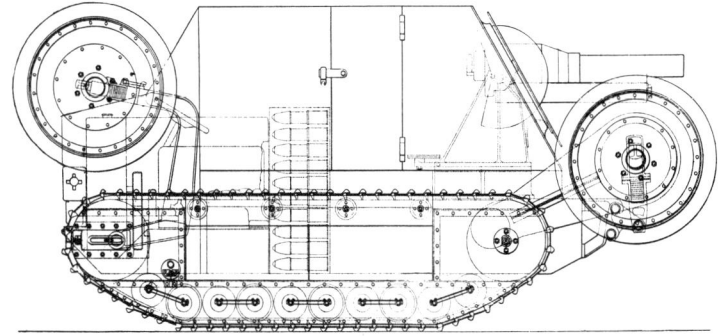

181

But at the same time as advanced SP guns appeared, the old types still persisted. St. Chamond brought out a horse-drawn 120 mm heavy field gun that had all the characteristics of an earlier epoch. From the same stable appeared the 185 mm barrel which was to have a long life.

although it is difficult to imagine how it would fare under bombardment. All in all, most chiefs of staff found themselves in the situation of town-planners responsible for the rational planning of a district without permission to demolish even a minimum of old buildings.

The war would take care of this on their behalf. And this was to be a war prepared by dictators who,

alone, could afford the luxury of properly modernizing their armies. The totalitarian states had waited until the tank and the bomber were fully developed before starting to produce them in series. Moreover, their workshops had developed entirely new types of guns for use in tanks and aircraft.

While the democracies were putting some of their old guns on to wheels with pneumatic tyres, the National-Socialist government had brought out an entirely new weapon, the 88 mm anti-aircraft gun. Unveiled and put to the test during the Spanish Civil War, some British military observers were astonished to see a weapon of such quality thus prematurely disclosed. Its performance was still unmatched in 1945.

But democracies do not always lag behind in matters of armaments. For instance, the Swiss High Command displayed a good sense of timing in equipping its entire infantry with a 47 mm gun for which top-quality ammunition was produced.

As there was no time to produce tanks, this small country bristled with "47"s. This move caused the German General Guderian some alarm, none of his tanks in 1939 being proof against such a projectile.

It is undoubtedly difficult to integrate artillery within a budget and a plan of reorganization when the increasing speed of technological progress is taken into account. For this reason, artillery remains

Right: 105 mm German field howitzer. The range at maximum elevation (40°) enabled it to send a 33 lb shell 5.5 miles. Left: 210 mm howitzer (field mortar) with an elevation of 45°.

The German 77 mm gun of 1935. The jacketed barrel was held on to the carriage by three guide clamps (see inset). The last example of its type, it was notable for its extreme simplicity.

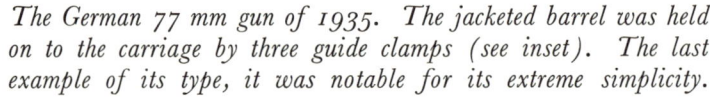

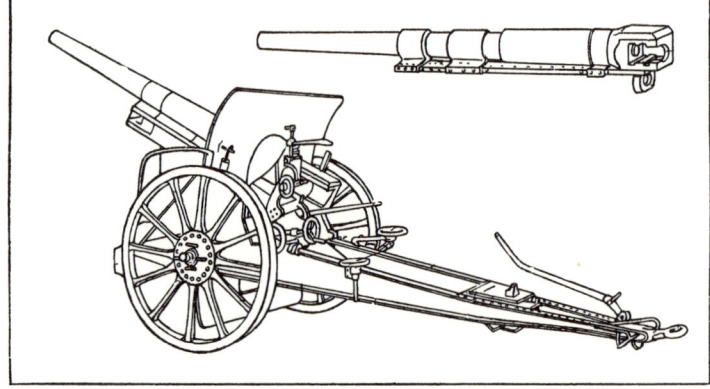

an art within which original solutions are always able to be developed.

But in all forms of art worthy of the name, errors are costly, as the following example illustrates. Guderian had requested a 50 mm long gun to be fitted to the Mark III tank. His proposition was accepted by all the departments concerned except one where an expert thought it would be a good idea to cut the barrel almost flush with the turret. Guderian explained how this mutilation came about. The expert in question had deemed it necessary so that the tank could swing its turret round in forests, thereby avoiding the trees!

The expert's decision was carried out and the result was disastrous. In use, the short barrel was totally ineffectual against the Soviet T 34 tanks whereas the originally specified long barrel would have been an efficient match to these new monsters. We now know that good, self-propelled artillery ought to have partially compensated for this error but, at the outbreak of war, assault artillery was in an embryonic state.

This digression has been necessary to illustrate the development of artillery during the uneasy truce that followed the First World War. Overburdened by a too heavily-stocked weapons park, artillery had allowed itself to be swamped by tanks, anti-tank and anti-aircraft guns and had not opted soon enough for a modern assault gun.

Lacking the financial resources, artillery was only semi-modernized by the introduction of some secondary improvements such as split-trail carriages enabling faster gun-laying in its lateral deviation and refinements to the sighting equipment. Apart from the "88", there were few really new guns. All ideas seemed to revolve around the old, some of which were sunk in concrete emplacements which cost three times less than gun-carriages fitted with tracks. Finally, motor traction was further developed. This was a spectacular step forward and tyres did not cost so much.

As for fire control, centralized control, practised with such success in the navy for over a generation, was still unknown to the gunners on land. It took the war for them to discover its necessity.

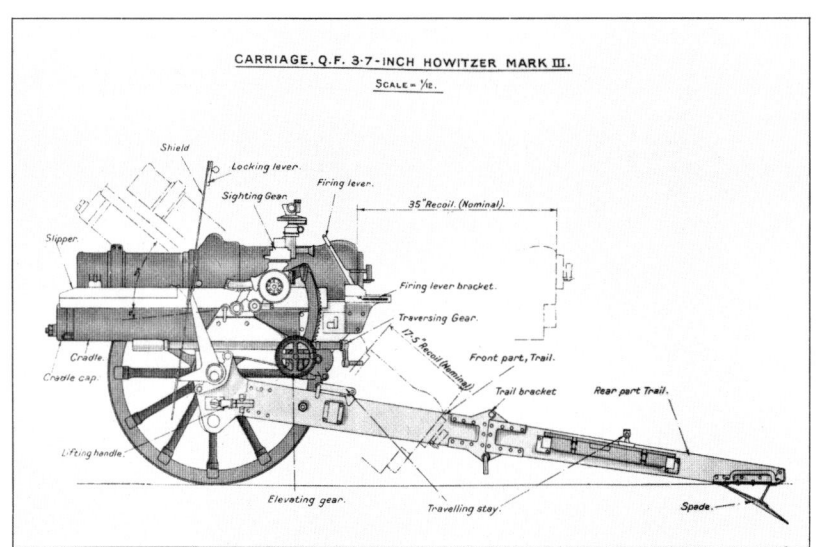

The British 3.7 inch howitzer of 1930. This simple and robust field piece could be manufactured in large numbers when needed.

THE SECOND WORLD WAR

The Second World War displayed very much more variety in its forms of combat than the first. Artillery had to adapt itself to widely differing conditions of use, making it difficult to summarize its history in a true light. The following pages will merely highlight those characteristic elements which serve as landmarks in such a wide field.

THE FALSE COMPETITION OF THE DIVE BOMBER

In a war of movement, targets never remain in the same position for a long time. It is therefore necessary to attack them quickly with heavy fire. In order to concentrate heavy fire from dispersed batteries, centralized fire control is essential. This was, in fact, carried out by aircraft control centres with which contact was easily established through the powerful radio available.

This type of control could not be achieved as rapidly as it should have been by the artillery. The laying of its telephone line network, mostly on the ground, took time. As for the artillery's radio system on medium wavelengths, it was not yet able to take the place of the telephone cable, proving unreliable, especially over undulating ground. Furthermore, particularly in those cases when the attack formations were moving further and further away from the firing positions, radio communications often failed when they were most needed.

Radio communication between armoured vehicles was much more dependable because of the shorter ranges, tanks mounting their attacks in close squadron formation. Therefore the armoured vehicles, in common with aircraft, had advantages over the artillery, preventing it from taking an integral part in the combined battle formation.

It is not surprising that the bomber and the tank, with their more convenient deployment, were preferred to the artillery. At the outbreak of all wars, the High Command of the military services are always tempted to use the most powerful means at their disposal in the first instance. In a similar way to chess players who do not fully understand the secret of the gambit, they overlook the actual capabilities of all their resources and unwittingly invite heavy rebuffs.

Medium (170 mm) German trench-mortar. It weighed 1,292 lbs, had a maximum range of 1.2 miles and fired a projectile of 120 lbs. The barrel was rifled, and it could be pivoted on its base plate. It is a lineal descendant of the mortars of the First World War, and was designed to be fired from trenches or dugouts (see inset).

Apart from the theorizing, the following paradoxical example of a battle without artillery support reveals its true function with the utmost clarity.

At the beginning of June 1940, the 38th German army corps, commanded by the future Field Marshal von Manstein, crossed the Somme west of Amiens.

One of the leading regiments encountered fierce defensive action by the French III/60 battalion. One German company was practically annihilated by a machine-gun section firing from the north corner of the Ailly wood. The German attack was halted in its tracks.

In order to prepare for another attack, the German divisional commander ordered two waves of twelve *Stukas* to bomb the corner of the wood. Under cover of the air raid, the attack was mounted but pushed back again. Although for the most part wounded, the machine gunners in the Ailly wood carried on firing.

The fact is that the pilots of the *Stukas*, dropping their bombs at noon in bright sunshine, mistook the edge of a shadow thrown by the northern edge of a timber wood with the actual corner of the forest. The German artillery could not intervene as communications with them had been broken off. Instead of waiting to re-establish liaison with the artillery, another attempt to attack was made with the help of mortars. The attack failed once more.

French 105 mm howitzer of 1939. The wheels moved out with the divided trail, forming an additional gunshield. Although it has an air of improvisation, this gun in fact was the forerunner of many other designs which adopted some of its features.

Believing the wood to be held by much greater forces than was actually the case, after three days of hard fighting two German divisions surrounded the Ailly wood. Nevertheless, the French battalion, extremely well led, escaped their clutches.

Because of the necessity of forgoing artillery support and of relying too much on the capabilities of the dive bomber, this regiment lost more than 15 % of its strength in killed and seriously wounded, heavy losses for a small operation.

Of course, this example should not lead to a generalization on the shortcomings of one arm in battle. In many cases, artillery was able to fulfil its basic functions but it is, nevertheless, true that the lack of reliable and rapid communications constituted one of its greatest handicaps.

The development of the fire plan was one method of surmounting this disadvantage. Map references of probable targets, pinpointed landmarks, the fringes of forests or other places which could be used as cover were all worked out in advance so that tasks could be fired effectively with the most elementary signals, such as Verey lights, or through the intermediary of the communications used by other services.

This was one of the great steps forward that came about in the Second World War. Another was the firing of concentrations, which could be termed the opposites to barrages. By making use of these vastly improved communications, it became possible to bring to bear anything from a troop to the entire Corps artillery on to one shoot, with devastating

A detachment of the RHA arriving at the gallop. This picture was taken in November 1939, during the "Phoney War".

185

Far East—1943. No matter what the theatre of war— two main considerations had to be taken into account by designers—the relation between punch and lightness and mobility.

Tunisia, 1943. In the foreground is a German field piece abandoned before the Allies' advance. Its double carriage helped in turning the weapon, and tyres its road speed. In the background is a British gun quad and limber.

The Ruhr in 1945—a typical scene from the close of the Second △ World War. The Americans have captured a 75 mm German anti-tank gun and turned it on its former owners. Such high-powered guns were useful in hand-to-hand combats.

Africa 1943. The German 88 mm was originally designed as a ▽ high-velocity AA gun, but Rommel found it the most effective anti-tank gun available. Its performance completely outclassed the tank-mounted guns of the time.

Russia 1943. As the war became prolonged, the elderly heavies were pressed into service again. This 1916 305 mm Skoda mortar is at maximum elevation. Falling from a great height, the shells reached supersonic speeds.

Russia 1943—the field of fire for this German 150 mm howitzer corresponds to a divisional front. In the winter of '43, the German industrial effort was turned towards the priority manufacture of munitions for the artillery.

An American 203 mm howitzer in France in 1944. The features of this gun were: maximum range 10.5 miles, effective range 7 miles, field of fire 60°, elevation from 1° to 63°, muzzle velocity 1,952 ft/seconds, road weight 13 tons, weight of projectile 198 lbs. The Anglo-Saxons always equipped their expeditionary forces with a strong artillery. For the invading Allies, weight played a smaller part than for the continental powers, whose communication systems was weaker than the US-UK systems.

effect. This was further expedited by the use of code words such as "Mike Target" "Uncle Target" to order various different concentrations. A variation of this was the linear concentration, called a "Stonk", which was used to saturate, for example, a length of wadi, a road, railway cutting or similar extended linear target.

Thus, by the end of the war, because of the American improvements to radio equipment, such as the VHF with fixed quartz frequencies, the artillery attached to an army unit could bring the fire of all its guns into effect nearly as quickly as the turrets of a battleship.

A US 155 mm gun, known as a "Long Tom", in France in 1945. This gun returned to a similar barrel to that developed by the French in 1920. Its long range of 14 miles gave it great flexibility in use.

France 1944—a US 203 mm gun. The more the war dragged on, the more sparing it was necessary to be with lives, and guns were made to do the work of battle. Ten Artillery groups fired in a day the equivalent of a kiloton of explosive.

The Shells

Shells could be fitted with mechanical time fuzes, a type of watch movement enabling the time of explosion during the trajectory to be determined to the tenth of a second. Nose fuzes in shells were better than the old graduated heads in which the explosion was caused by the less accurate pyro-technical method, thus greatly reducing the fuze timing. In other words, firing with time fuzes using this new material enabled the dug-in enemy to be bombarded with twice as much shell-fire.

Just after the invasion of Normandy by the Allies, the Germans launched their VI, or flying bomb, attack on London. This weapon was so destructive that in a short time, more damage was done to London than had been caused up to then by the conventional bombing. At first, the RAF was charged with the destruction of the VIs over the Channel and inland, up to a distance approximately halfway to London. However, it was soon realized that the simple guiding mechanism of this weapon met the basic presumption of anti-aircraft predictors i.e. constant course, speed, and height. The most massive deployment of anti-aircraft artillery ever known was ranged along the South Coast of England in the path of the approaching missiles. Then followed a day and night engagement, which

Normandy 1944 - a 76.2 anti-tank gun passes through a damaged village. Although the heavy-gun problems had not yet been solved, light guns had great value.

The battle of Khorson, Russia, 1944, by Krivonogov. A graphic illustration of what could happen to Artillery that was forced to use the roads by the nature of the surrounding terrain, and was caught before it had time to unlimber its guns.

Holland 1944—this German gun has automatic ammunition feed, recoil compensators, and a ranging table.

Russia 1944—Soviet artillery pushing on over difficult country. This is a clear cut case for the need for light and manoeuvrable guns.

lasted for weeks on end, in which the men and women of Anti-Aircraft Command under General Sir Frederick (Tim) Pile had no respite, until the advancing British 2nd Army overran the launching platforms. The accuracy of the anti-aircraft fire was such that there was an immediate enquiry if a V1, or to give it its code name, "Diver", passed through the defences. This showed that the opinion expressed by General Foy in the Napoleonic wars was still very much to the point.

This prolonged action was noteworthy in several respects: that women were in action not only in the control rooms, but also actually on the gun positions, operating fire control instruments; that like the 1940 Battle of Britain, when the Royal Air Force stood between the Luftwaffe and capitulation, so, had the gunners failed to have stopped the V1s, London would have faced certain destruction, with all that that might have entailed.

There was another parallel between the Battle of Britain and the Diver battle. This was the use of radar, which was first harnessed by Sir Robert Watson-Watt in 1935. During the Battle of Britain, radar signals were sent out from 300 ft and 400 ft high masts, erected in groups of 4 around the English coasts to detect approaching aircraft. The position and height of the enemy was then passed

to the fighter planes, who were "scrambled" to intercept. During the Diver battle, radar was used automatically to follow the target and to pass information to the predictors. Quite apart from this, in a very minituarised version it was to play a vital role in the action.

Months earlier, British scientists had succeeded in reducing the size of a radar set to such a degree that a tiny version could be used as a fuze in the nose of a shell. It was felt that the USA had the better potential for production, and after discussion in London with US scientists, the USA accepted the task on behalf of the Allies. These fuzes were ready in quantity in time for Diver. They eliminated the necessity for fuze setting, as they exploded the shell when a radar signal was reflected by the approaching missile. This fuze was known as the Proximity or Variable Time (VT) fuze, and was used again by the US field artillery in the Ardennes and by the British in the Rhine Crossing. The air burst effect at a constant height above the ground no matter what the ground contour was devastating. Probably the most effective combination of equipment during Diver was the US-made SCR 584 radar with the British no. 11 Predictor, served by the Radar and Visual Tracker controlling the QF 3.7 inch anti-aircraft gun firing HE shells with VT fuzes. This was the peak of anti-aircraft efficiency, and probably the last great anti-aircraft action.

On the subject of shells, the nebulogenes or tear gases are worthy of note. The Anglo-Saxons, in particular, were to make quite extensive use of them. In order to understand the effectiveness of this type of shell, it is only necessary to think of what might have happened at the battle of Ailly wood if the French machine-gunners could have been blinded for a critical moment during the German advance.

The Barrels

Known as *autofrettage* or self-shrinking, a method consisting of creating a considerable inside pressure within the barrel endows it with exceptional strength. In order to create this pressure, the cannon treated in this way is fired with heavy charges.

This method is very much less costly than that of inserting a first tube into a second. In fact, in order to make these "sleeved" barrels, the enveloping part must first be heated so that it expands sufficiently to be able to slide over the first tube while the second one is still red-hot. The new process henceforth enabled all troops to be assured of high performance weapons such as had been hitherto reserved for the Navy.

The guns of the Second World War, produced with the improvements of this process, easily lasted for 5-6,000 rounds in the case of heavy guns and twice as much for howitzers. Their length of life also logically depended on the quality of the projectile and the type of driving-bands on the shells. Lacking copper at one time, the Russians made their shells with less malleable bands which decreased the life of the barrels by half.

The old and glorious "75's", their horse-drawn batteries often severely mauled by the *Stukas* on the Continental battle-fields, came back into their own in North Africa. The Officer Commanding the Afrika Korps had a healthy respect for them in the hands of the defenders of Bir Hakim.

The 105 calibre slowly came into general use in most armies. Curiously enough, the barrel of its howitzer proved almost immune to wear although that of the cannon was far less indestructible.

1944—Soviet heavy guns firing on the Germans besieging Leningrad. The battery consists of 152 mm howitzers.

A French 155 mm SP howitzer, mounted on an AMX chassis. In its firing position, the vehicle is steadied by the two spade trails. It is not heavily armoured, for these guns do not need to stay long in any one position.

The American M-109 is a development of the 155 mm howitzer. It has been designed to withstand all forms of warfare, Atomic, Bacteriological, and Chemical (ABC). Constructed in large numbers, it is standard equipment for many Western European countries.

Practically a mobile fortress, this Swedish Bofors 155 mm piece is a gun and not a howitzer. Its original loading system enables it to fire a score of rounds a minute. Little by little the 105 mm calibre was abandoned, for the shells lacked the power to penetrate the armour of modern tanks.

A post-war exercise—"Bourdon II"—of the French 35th Parachute Regiment. The helicopter has become the army pack-horse, and is well equiped to transport guns over difficult country.

A US 75 mm recoilless gun, introduced at the end of the war. Light and mobile, with a weight of 114 lbs and a useful range of 3.7 miles, it had much to commend it. But its big disadvantage was the very heavy ammunition which when fired left a large cloud of smoke, making concealment difficult.

A 105 mm Italian mountain piece. Throughout history, the Italians have been known for their guns. The weapon breaks down into eleven loads for easy transport.

For their part, the Russians opted for a calibre of 122 mm and very high charges, giving their shells a near-supersonic speed. As the enemy could not hear the approach of the first salvo, they were not warned by the whistling of shells as they were during shell-fire at subsonic speeds.

Nevertheless, trajectories at very high speeds are relatively flat and do not lend themselves to use in mountainous and broken country. Therefore, in all armies, weaker charges were developed so that the shells would make a more curved trajectory, thus enabling more efficient fire to be directed on the blind slope of a ridge.

Pieces of ordnance, such as mortars, firing at angles greater than 45° were constructed to ensure infantry support in the most undulating terrain. The Americans adopted the excellent 155 calibre gun which the French had perfected during the previous war. This piece of ordnance, of European origin, reappeared on the Continent with Eisenhower's troops. With the 203 mm pieces, it constituted the backbone of the American artillery.

It is unnecessary to enumerate all the related equipment but Dora is worthy of mention. Dora was the daughter or, perhaps, niece of Big Bertha. Her 800 mm calibre shell enabled her to bring down the heaviest fortifications of Sebastopol. A considerable research and construction programme had been carried out to perfect her and many German military voices were raised in protest against such uselessness. In order to bring her into position, it was necessary to employ more than five thousand "servants" and ten parallel railway tracks.

Dora's shells were effective in bringing down the fortress of Maxim Gorki in 1942. On the other hand, they were of no use against the stubbornly defended ruins of Stalingrad since Dora's weight prevented her from being transported so far. Her barrel of 80 cms in calibre recalls the ancient heavy bombards with which artillery made its first appearance on the pages of history. This strange weapon represented the end of an era; the beginning of the next was heralded by the large-scale introduction of the heavy water bomb which was being developed as rapidly as possible by both sides.

The biggest modern guns are the direct descendents of the old siege guns. This 203 mm SP gun is in the classical tradition, placing at the disposal of an army commander a concentrated mass of heavy fire. (above)

This 175 mm American SP gun is useful to a divisional commander with a large front to control. It has a range of more than 15.5 miles, and is a very accurate weapon, its shells landing in a very small aera at long ranges. (below)

CONSIDERATIONS ON THE USE OF ARTILLERY AT THE END OF THE SECOND WORLD WAR

At the end of this conflict, artillery was again perfectly integrated into the combined operations force in the sense that it fulfilled specific functions for which other arms were not equipped.

Manoeuvrability

When the tanks of the 2nd Armoured Division asked for its support in Alsace or the Black Forest, it was only necessary to transmit the coordinates and one code word such as "Piledriver", "Hammer" or "Anvil" to bring about the type of fire wished at the desired place in the shortest possible time.

Equally Efficient for Siege Operations

When General Montgomery decided to clear a path to allow his troops to cross a mine-field consisting of some 500,000 unmapped mines in front of El Alamein, 1,000 field pieces each fired nearly a 1,000 shells which blew up the hidden mines.

When Stavka broke through the Vistula front in January 1945, he resorted to the method perfected by Bruchmüller[1], famous for his *Trommelfeur* in 1918.

Tracks, radio, cartography, telemetry, aerial photography, radar, as well as progress in the study of topography and meteorology have all contributed to the uneasy integration of artillery. The necessary qualities for manoeuvres in field and static operations had been combined when before they had seemed hardly compatible.

This war also saw guns taking to the air. In accordance with the need for artillery to be in close support of assaulting troops, a natural development was that of self-propelled artillery on tracks, that could quickly be brought to bear on any obstacles encountered by the assaulting troops. Where greater distances were involved, airborne and parachute artillery were developed to support the airborne troops in deep penetration operations.

However, no study of artillery would be complete without taking into account the contribution of the rocket and the appearance of the atomic weapon, but lack of space prevents more than a passing reference to be given to them.

* * *

[1] Oberst Georg Bruchmüller: *Die Artillerie beim Angriff im Stellungs-Krieg*. Charlottenburg 1926.

A US 280 mm atomic gun, which can be considered as a transitional point in the ▷ development of artillery. The atomic SP gun is probably the rising fashion in weapons.

CONCLUSION

After Hiroshima and Nagasaki, the knowledge that a new weapon of undreamt-of potentiality and effect started a world-wide re-evaluation of weapon systems. Perhaps it was not just the scale of the destruction caused by the atomic bomb that caused the concern, for comparable devastation had been seen after the bombing of Dresden by the Allied Air Forces, and thousands of lives had been lost in the artillery bombardments on the Somme in the First War. Equally, the V2 rockets had been almost as impossible to destroy after launching with the weapons and equipment then available. The true foundation of the myth of the atomic weapon was its compactness, enabling a gigantic destructive force to be contained in a comparatively small container, which could be delivered either by rocket, aircraft or by guns rapidly and accurately.

However, the possession of this potential posed its own problems. Bombs dropped from aircraft tend to be indiscriminate in effect, and although political reasons might require a city or a province to be obliterated, military requirements would perhaps call for rather more limited, concentrated effects. The same reasons apply to the larger rocket systems to a lesser degree. These do not, however, apply to the atomic shell, fired from a gun, where the comparitively small atomic warhead, although containing a very large destructive power in relation to size, can be considered as limited in effect, can be directed accurately, and its effects can be predicted. Atomic cannon and infantry missiles may be considered as strictly military weapons, which will find their place on future battle-fields when fear of reprisals makes the use of megaton missiles unlikely.

Despite uninformed opinions to the contrary, it is improbable that military reasons would dictate the use of the megaton missiles. To render a country a desert would not serve any purely military end, and in any case, the so-called deterrent effect tends to cancel out the use of really large nuclear weapons. Atomic cannon, however, are likely to re-establish the gun as the most important weapon in the army commander's armoury. Thus the wheel turns, and in this century, we have seen military thought advance from the bayonet to the gun, from the gun to the tank and the aircraft, and again back to the gun.

The Bundeswehr's multiple rocket launcher is another modern development of an old idea. Its primary function is the rapid delivery of a heavy and accurate simultaneous fire.

Perhaps the giant Russian satellite carrier is really designed to launch a form of orbital artillery. It stands at the beginning of a new era of weapons.

198

DESCRIPTION OF A BRONZE CANNON

1. axle-tree
2. axle
3. elevating screw
4. trunnion
5. drag link
6. dolphins
7. cap-square
8. chase
9. muzzle
10. sponge
11. rammer
12. portfire cutter
13. shoe wheel-brake
14. lifting handle
15. trail-eye
16. traversing handspike
17. cascable
18. vent or touch-hole
19. bore

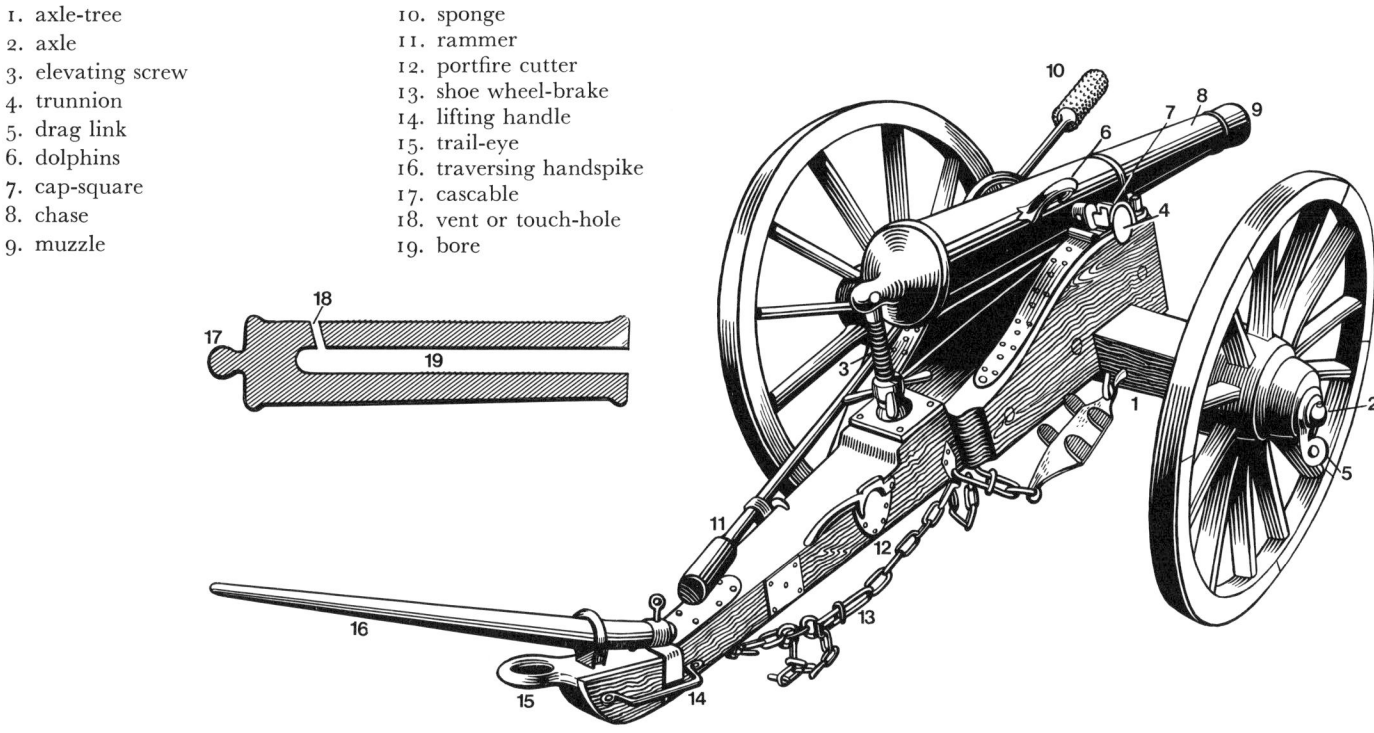

DESCRIPTION OF A GUN (75 mm FIELD-GUN)

1. barrel jacket
2. barrel
3. muzzle roller
4. gun shield
5. shield brace bracket
6. elevating sector
7. range scale
8. recoil mechanism
9. breech block
10. operating handle (breech)
11. breech
12. aiming gear
13. firing hammer
14. eccentric breech closing screw
15. piston rod cushion
16. traversing handwheel
17. travelling lug block
18. pawl lever (recoil travel lock)
19. tiere rods
20. piston rod coupler (recoil brake)
21. elevating yoke
22. rocker
23. layer's seat
24. brake shoe
25. axle spade
26. right trail handle
27. upper trail plate
28. trail
29. lunette
30. spade
31. trail
32. tie rod shackle
33. knee guard
34. recoil mechanism
35. lanyard handle, firing hammer

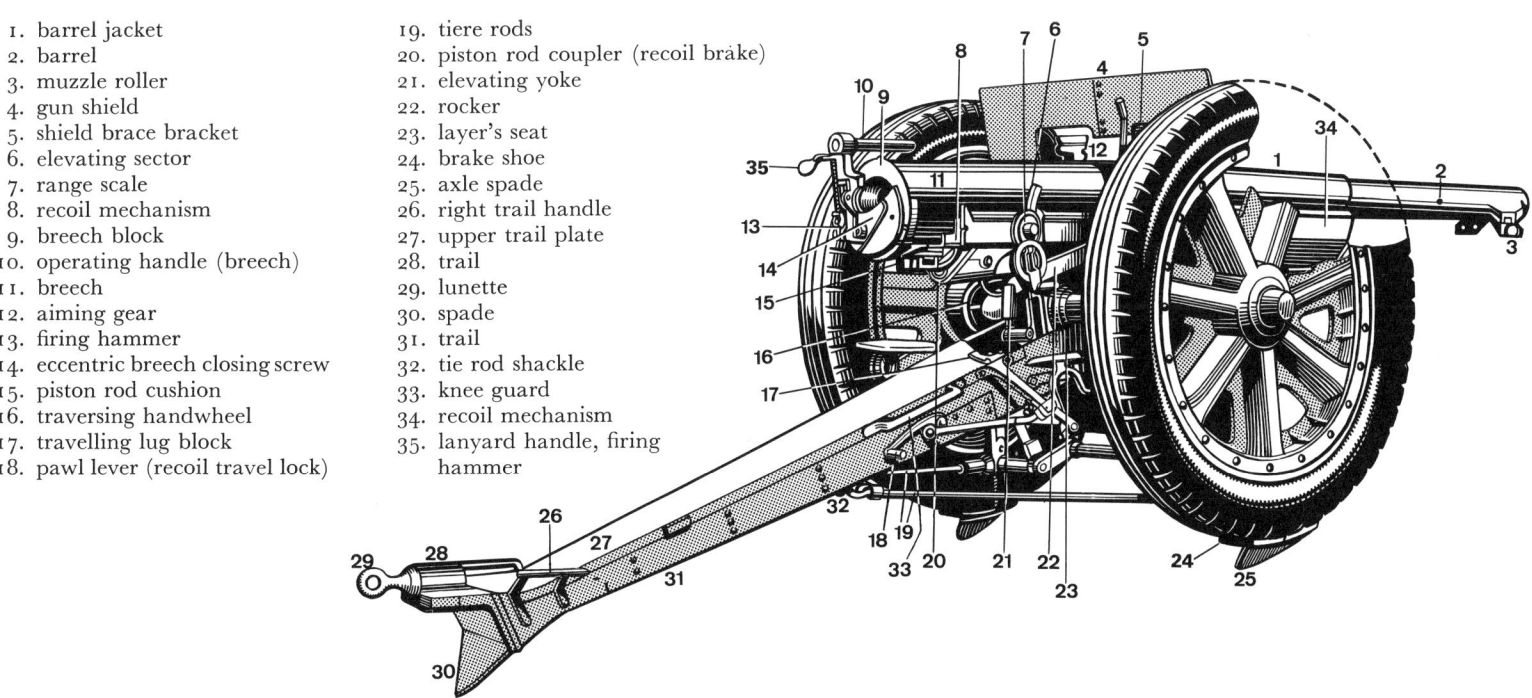

199

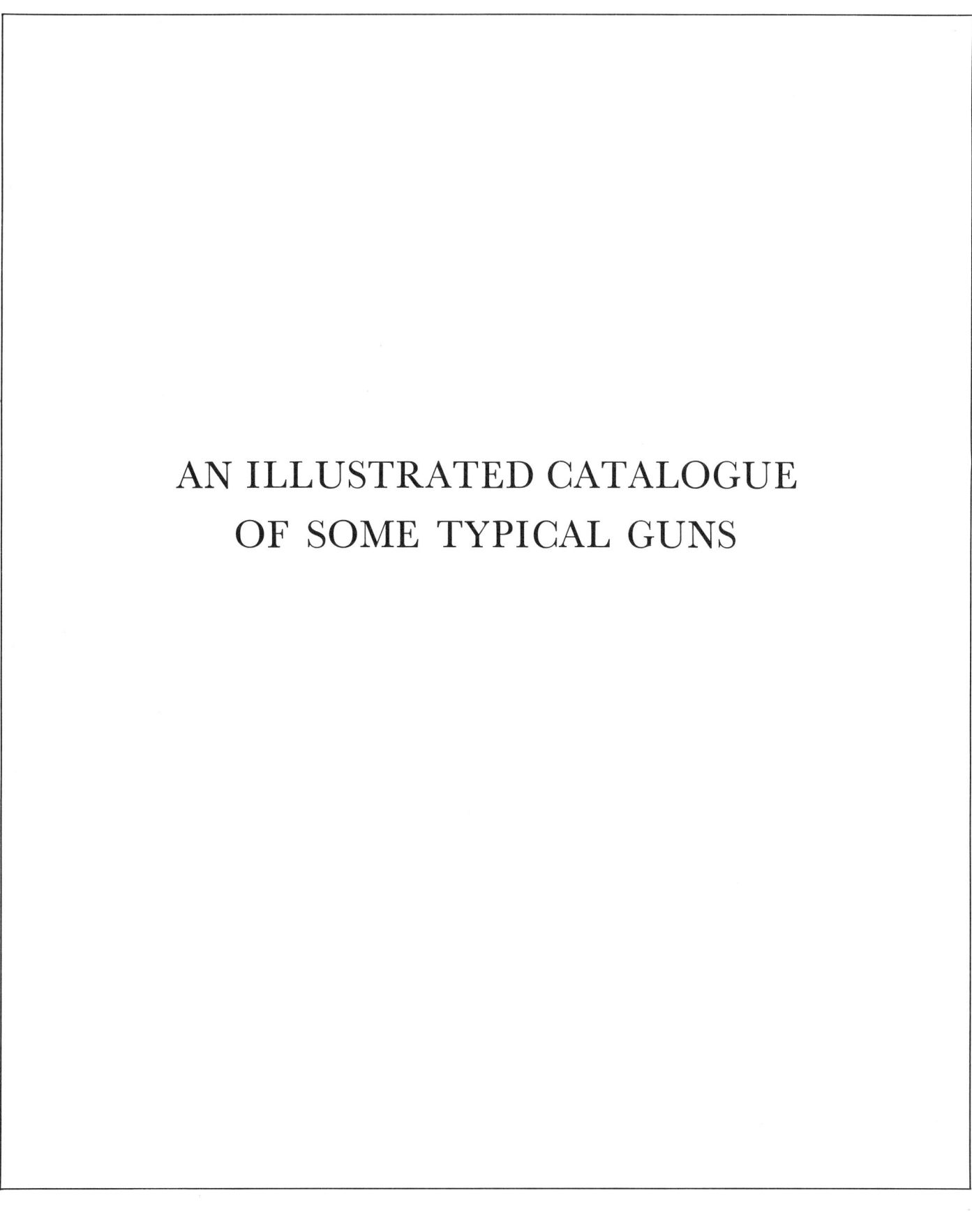

AN ILLUSTRATED CATALOGUE
OF SOME TYPICAL GUNS

Type: Perrier
Date: 2nd half of fourteenth century
Material: wrought iron
Calibre: 6.4 ins
Length of barrel: 3 ft
Weight: 275 lbs
Where kept: Heeresgeschichtliches Museum, Vienna

Type: Italian perrier
Date: 2nd half of fourteenth century
Where kept: Bernisches Historisches Museum, Berne
Remarks: Carriage reconstructed after a manuscript of
Marianus Jacobus of Siena

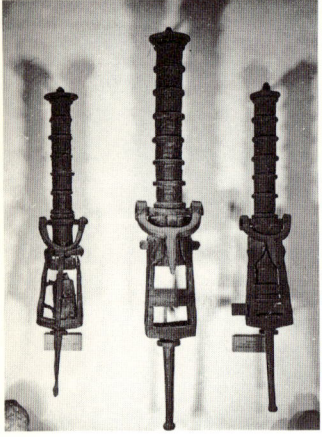

Type: Cannon
made of wrought iron
Date: End of fourteenth
century, beginning of
fifteenth century

Type: Giant perrier
Name of gun: Pumhardt
Date: first half of fifteenth century
Calibre: 35.2 ins
Length of barrel: 8 ft 3 ins
Weight: 7 tons 1,920 lbs
Where kept: Heeresgeschichtliches Museum, Vienna

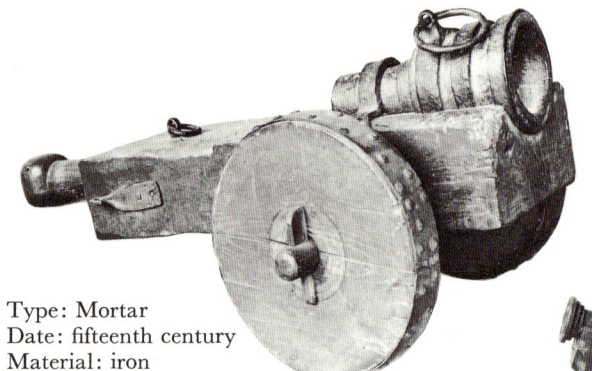

Type: Mortar
Date: fifteenth century
Material: iron
Calibre: 6.8 ins
Length of barrel: 1 ft 3 ins
Weight: 81 lbs
Where kept: Bernisches Historisches Museum, Berne
Remarks: Carriage reconstructed

Type: Perrier
Date: first half of fifteenth century
Material: wrought iron
Calibre: 6.4 ins
Length of barrel: 3 ft
Where kept: Heeresgeschichtliches Museum, Vienna

Type: Culverin
Date: first half of fifteenth century
Material: wrought iron
Calibre: 1.4 ins
Length of barrel: 4 ft 6 ins
Where kept: Musée de Morat

Type: Culverin
Date: c. 1450
Material: wrought iron
Calibre: 5.6 ins
Length of barrel: 2 ft
Origins: Burgundy
Where kept: Musée de Morat

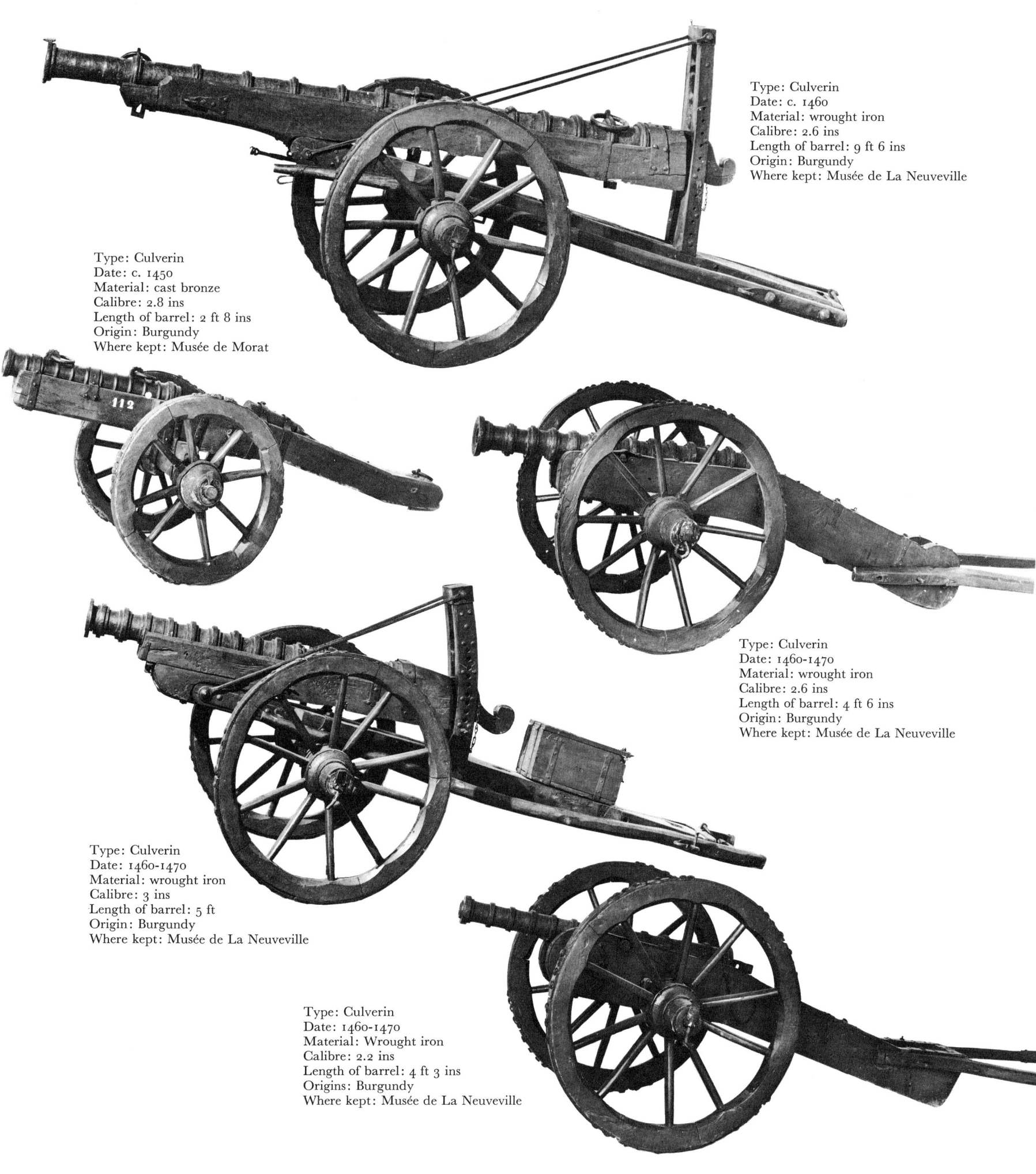

Type: Culverin
Date: c. 1460
Material: wrought iron
Calibre: 2.6 ins
Length of barrel: 9 ft 6 ins
Origin: Burgundy
Where kept: Musée de La Neuveville

Type: Culverin
Date: c. 1450
Material: cast bronze
Calibre: 2.8 ins
Length of barrel: 2 ft 8 ins
Origin: Burgundy
Where kept: Musée de Morat

Type: Culverin
Date: 1460-1470
Material: wrought iron
Calibre: 2.6 ins
Length of barrel: 4 ft 6 ins
Origin: Burgundy
Where kept: Musée de La Neuveville

Type: Culverin
Date: 1460-1470
Material: wrought iron
Calibre: 3 ins
Length of barrel: 5 ft
Origin: Burgundy
Where kept: Musée de La Neuveville

Type: Culverin
Date: 1460-1470
Material: Wrought iron
Calibre: 2.2 ins
Length of barrel: 4 ft 3 ins
Origins: Burgundy
Where kept: Musée de La Neuveville

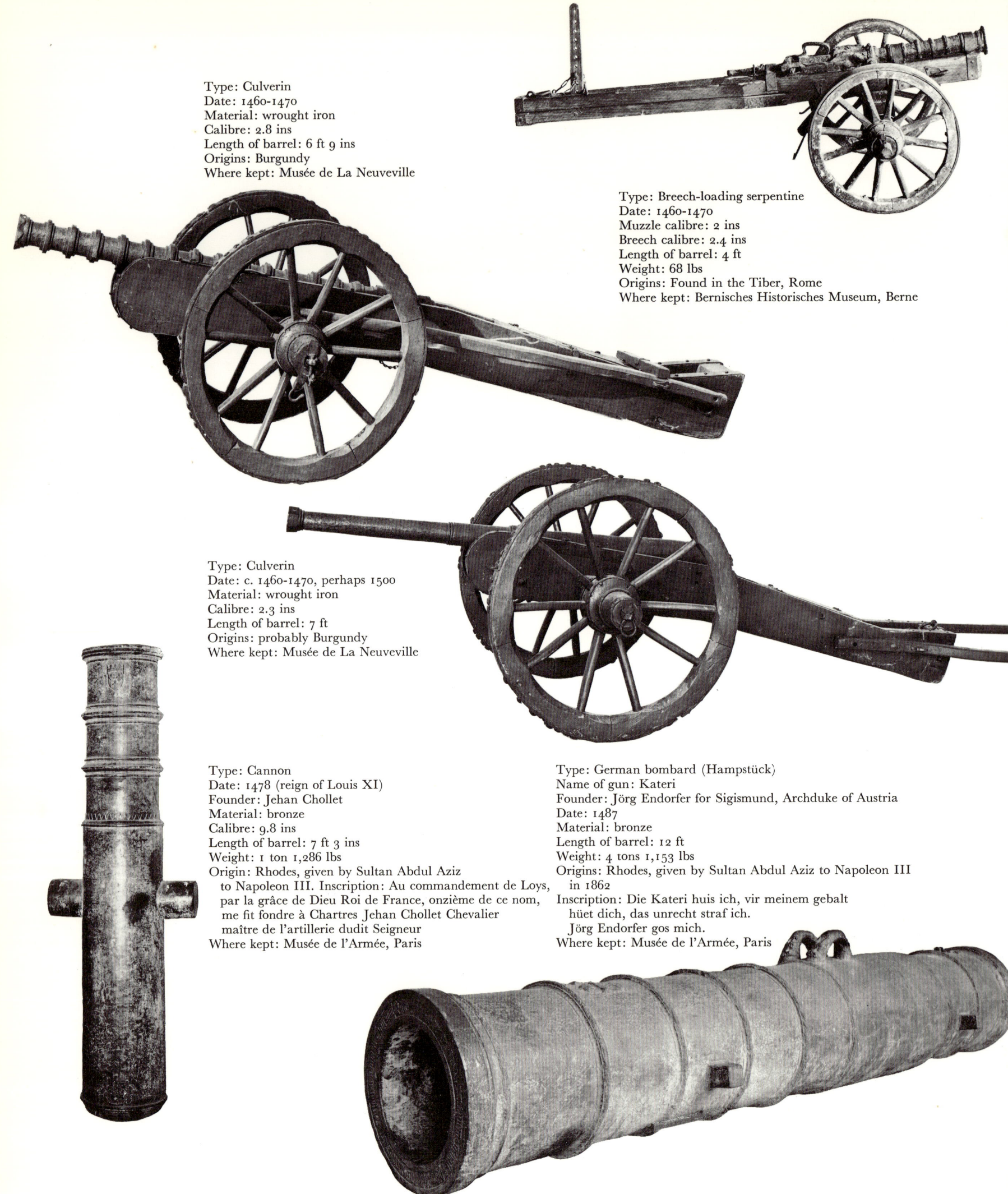

Type: Culverin
Date: 1460-1470
Material: wrought iron
Calibre: 2.8 ins
Length of barrel: 6 ft 9 ins
Origins: Burgundy
Where kept: Musée de La Neuveville

Type: Breech-loading serpentine
Date: 1460-1470
Muzzle calibre: 2 ins
Breech calibre: 2.4 ins
Length of barrel: 4 ft
Weight: 68 lbs
Origins: Found in the Tiber, Rome
Where kept: Bernisches Historisches Museum, Berne

Type: Culverin
Date: c. 1460-1470, perhaps 1500
Material: wrought iron
Calibre: 2.3 ins
Length of barrel: 7 ft
Origins: probably Burgundy
Where kept: Musée de La Neuveville

Type: Cannon
Date: 1478 (reign of Louis XI)
Founder: Jehan Chollet
Material: bronze
Calibre: 9.8 ins
Length of barrel: 7 ft 3 ins
Weight: 1 ton 1,286 lbs
Origin: Rhodes, given by Sultan Abdul Aziz
  to Napoleon III. Inscription: Au commandement de Loys,
  par la grâce de Dieu Roi de France, onzième de ce nom,
  me fit fondre à Chartres Jehan Chollet Chevalier
  maître de l'artillerie dudit Seigneur
Where kept: Musée de l'Armée, Paris

Type: German bombard (Hampstück)
Name of gun: Kateri
Founder: Jörg Endorfer for Sigismund, Archduke of Austria
Date: 1487
Material: bronze
Length of barrel: 12 ft
Weight: 4 tons 1,153 lbs
Origins: Rhodes, given by Sultan Abdul Aziz to Napoleon III
  in 1862
Inscription: Die Kateri huis ich, vir meinem gebalt
  hüet dich, das unrecht straf ich.
  Jörg Endorfer gos mich.
Where kept: Musée de l'Armée, Paris

Type: Bombard
Date: End of fifteenth century
Founder: made by order of Grand Master Pierre d'Aubusson,
    of the Order of St John, and kept in Rhodes
Material: bronze
Calibre: 23.2 ins
Length of barrel: 6 ft 3 ins
Weight of shot: 574 lbs (granite)
Weight: 3 tons 595 lbs
Inscription: F PETRUS DAUBUSSON M HOSPITALIS IHER
Where kept: Musée de l'Armée, Paris
Remarks: The slots at both ends of the piece were for levers
    used to screw and unscrew the chamber from the barrel.
    It was necessary to divide these giant pieces
    for transport

Type: Culverin
Date: c. 1490
Calibre: 2.2 ins
Length of barrel: 6 ft 3 ins
Weight: 374 lbs
Where kept: Bernisches Historisches Museum, Berne

Type: Tarrasbüchse
Date: fifteenth century
Material: wrought iron
Calibre: 1.8 ins
Length of barrel: 4 ft 9 ins
Weight: 206 lbs
Where kept: Heeresgeschichtliches Museum, Vienna

Type: 7 barrels mounted in parallel
Date: fifteenth – sixteenth centuries
Calibre: 1.2 ins
Length of barrels: 2 ft 6 ins – 3 ft 9 ins
Where kept: Heeresgeschichtliches Museum, Vienna

Type: Serpentine
Date: 1505
Calibre: 5.2 ins
Length of barrel: 14 ft 9 ins
Origin: Republic of Ragusa
Where kept: Heeresgeschichtliches Museum, Vienna

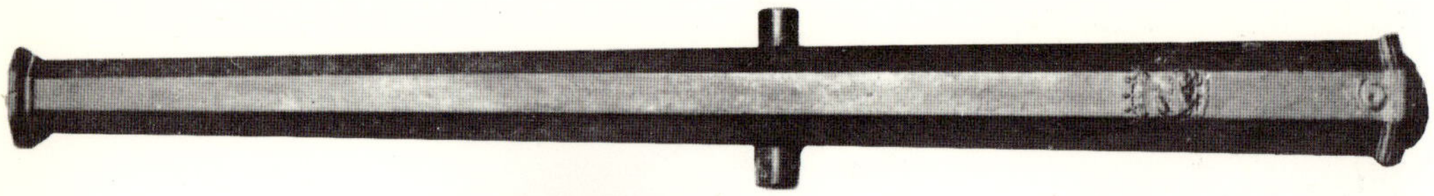

Type: Culverin
Date: reign of François I (1515-1547)
Material: bronze
Calibre: 3 ins
Length of barrel: 9 ft
Weight: 1,342 lbs
Where kept: Musée de l'Armée, Paris
Remarks: 8-sided barrel, bearing a crowned salamander,
the device of François I

Type: Double cannon,
    called Ehrenbreitstein culverin
Name of gun: Greif (Griffon)
Date: 1524
Founder: Simon, for the Electoral Prince
    Archbishop of Trêves, Richard von Greifenklau
Material: bronze
Calibre: 11.4 ins
Length of barrel: 16 ft
Weight: 12 tons 815 lbs
Origin: Frankfurt
Inscription: Der Greif heis ich
    minem genedigen Hern von Drir din ich
    wo er mich gewalden
    do will ich Dorn and Mauern zu spalten.
Where kept: Musée de l'Armée, Paris
Remarks: This cannon was made from pieces taken
    from the castle of Franz von Sickingen. Taken by
    the French at Ehrenbreitstein in 1799, it was lodged
    in the arsenal at Metz, and entered the museum in
    Paris in 1865. It was taken by the German in 1940,
    and returned to Paris in 1946

Type: Demi-kartaune
Name of gun: Amsel
Date: 1579
Founder: Martin Hisger
Calibre: 6.4 ins
Length of barrel: 10 ft 3 ins
Where kept: Heeresgeschichtliches Museum, Vienna

Type: Falcon
Date: reign of Henri II (1547-1559)
Calibre: 2.2 ins
Weight of shot: c. 1.5 lbs
Length of barrel: 6 ft 6 ins
Inscription: The fleur de lys, interlaced ciphers
    of Henri and Catherine, and the crescent of Diana,
    surrounded with arcs, on the superior surface
Where kept: Outside the church of St Louis,
    Invalides, Paris

Type: 6-pounder cannon (?)
Date: 1590
Material: Bronze
Calibre: 3.6 ins
Length of barrel: 9 ft
Inscription: Protector meus et refugium meum es tu. Claudius
    a Guisia abbas cluniacensis fieri fecit anno Domini 1590.
Arms of the Bastard Claude de Guise, Abbot of Cluny
Where kept: Musée de l'Armée, Paris

Type: Howitzer
Date: 1594
Founder: Hanns Dinckelmaier
Calibre: 5 ins
Length of barrel: 4 ft 6 ins
Where kept: Heeresgeschichtliches Museum, Vienna

Type: Serpentine
Date: 1628
Calibre: 1.5 ins
Length of barrel: 3 ft 6 ins
Where kept: Heeresgeschichtliches Museum, Vienna

Type: 6-pounder cannon (?)
Date: beginning of seventeenth century
Material: bronze
Calibre: 4.3 ins
Length of barrel: 10 ft 9 ins
Where kept: Musée de l'Armée, Paris
Remarks: Bears arms of Cardinal Richelieu. One of
six cannon kept in his private arsenal, and which were
displayed on the outer walls of his chateau.
Seized by the Revolutionary army in 1792

Type: Breech-loading cannon
Date: 1635
Calibre: 1.4 ins
Length of barrel: 8 ft
Where kept: Heeresgeschichtliches Museum, Vienna

Type: Lederkanone, or leather gun
Date: 1640
Calibre: 2.6 ins
Length of barrel: 9 ft 9 ins
Where kept: Heeresgeschichtliches Museum, Vienna

Type: Falcon
Date: 1682
Calibre: 2.1 ins
Length of barrel: 4 ft 9 ins
Where kept: Heeresgeschichtliches Museum, Vienna

Type: Mortar
Name of gun: Schwan (swan)
Date: 1704
Founder: Daniel Wyss
Material: Bronze
Calibre: 3.1 ins
Length of barrel: 8.4 ins
Weight: 15.6 lbs
Where kept: Bernisches Historisches Museum, Berne

Type: 4-pounder cannon
Date: 1716-1726
Founder: Abraham Gerber
Calibre: 3.3 ins
Length of barrel: 4 ft
Weight: 516 lbs
Where kept: Bernisches Historisches Museum, Berne
Remarks: Johann Rudolf Wurstemberger invented this
    quick-firing cannon, and was paid 2,500 crowns on
    delivery of 12 cannon of this type

Type: 100-pounder mortar
Date: 1754
Founder: Balthasar Herold
Calibre: 11.3 ins
Length of barrel: 1 ft 9 ins
Where kept: Heeresgeschichtliches Museum, Vienna

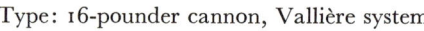

Type: 16-pounder cannon, Vallière system
Name of gun: Glaucus
Date: 1744
Founder: Jean Maritz
Inscription: Ultima ratio regum –
    Louis Charles, duc de Bourbon, duc d'Aumale –
    Nec pluribus impar
    (above the sun and the royal arms)
Where kept: Musée de l'Armée, Paris

Type: 16-pounder cannon, Vallière system
Name of gun: La Furibonde
Date: 1732
Where kept: Musée de l'Armée, Paris

Type: 2-pounder regimental cannon
Date: 1757
Founder: Samuel Maritz
Material: bronze
Calibre: 2.7 ins
Length of barrel: 4 ft
Weight: 352 lbs
Inscription: Sam Maritz FEC MDCCLVII
Where kept: Bernisches Historisches Museum, Berne

Type: 100-pounder mortar
Name of gun: Adler (eagle)
Date: 1764
Calibre: 10 ins
Length of barrel: 3 ft
Weight: 1,669 lbs
Where kept: Bernisches Historisches Museum, Berne

Type: 8-pounder cannon, Gribeauval system
Name of gun: Le Rigide
Date: 1789
Founder: J. Berenger
Material: bronze
Length of barrel: 6 ft 6 ins
Where kept: Musée de l'Armée, Paris

Type: 4-pounder cannon
Date: 1792
Material: bronze
Calibre: 3.4 ins
Length of barrel: 5 ft
Inscription: AN – Liberté – Egalité
Where kept: Musée de l'Armée, Paris
Remarks: AN designated National Arm
    (Arme Nationale) which were a municipal
    responsibility under the law of 3 August 1792

Type: 12-pounder cannon,
    Gribeauval system
Date: 1794
Founder: Perier Frères
Material: bronze
Calibre: 4.8 ins
Length of barrel: 7 ft 6 ins
Inscription: AN – Liberté – Egalité
Where kept: Musée de l'Armée, Paris

Type: Field gun
Name of gun: Brezin
Date: 1800
Material: brass
Calibre: c. 5.2 ins
Weight of shot: 18 lbs
Period of use: 1800-1815
Origins: Made in the Paris Arsenal, captured by
    the British at Waterloo in 1815
Where kept: Old College, Sandhurst,
    National Army Museum, Camberley

Type: 6-pounder cannon
Date: Not known
Length of barrel: 5 ft
Period of use: 1840-1846
Where kept: National Army Museum, Camberley

Type: Fortress piece
Date: 1859
Calibre: 6 ins
Length of barrel: 6 ft 9 ins
Weight: 1,936 lbs
Where kept: Heeresgeschichtliches Museum, Vienna

Type: Field gun
Date: 1915
Calibre: 6 ins
Length of barrel: 19 ft 6 ins
Weight: 4 tons 1556 lbs
Where kept: Heeresgeschichtliches Museum, Vienna

Type: Austrian field gun
Date: 1863
Calibre: 3.2 ins
Length of barrel: 4 ft 3 ins
Weight: 578 lbs
Where kept: Heeresgeschichtliches Museum, Vienna

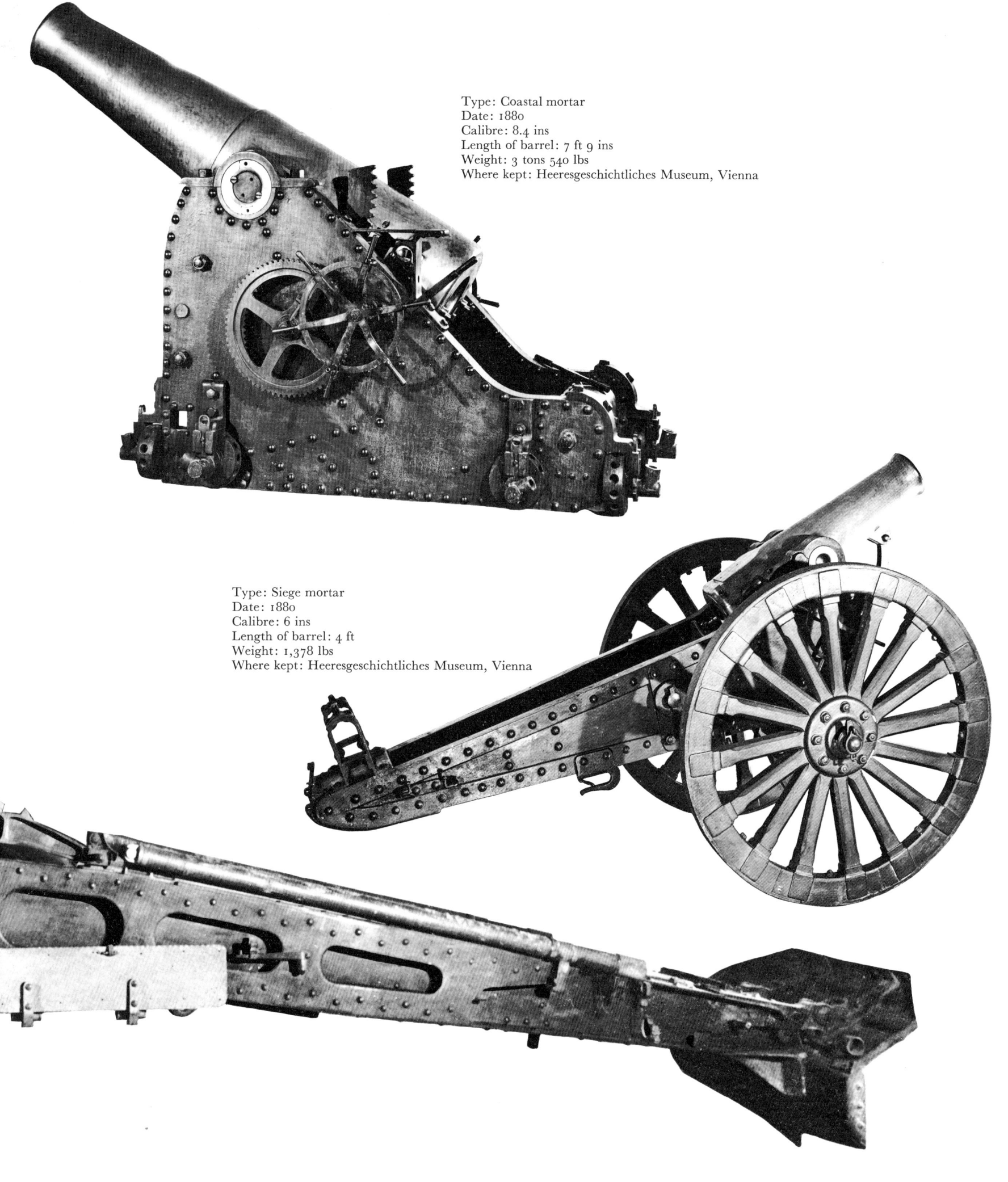

Type: Coastal mortar
Date: 1880
Calibre: 8.4 ins
Length of barrel: 7 ft 9 ins
Weight: 3 tons 540 lbs
Where kept: Heeresgeschichtliches Museum, Vienna

Type: Siege mortar
Date: 1880
Calibre: 6 ins
Length of barrel: 4 ft
Weight: 1,378 lbs
Where kept: Heeresgeschichtliches Museum, Vienna

Type: Howitzer
Date: 1916
Calibre: 15.2 ins
Length of barrel: 21 ft
Weight: 20 tons 740 lbs
Where kept: Heeresgeschichtliches Museum, Vienna

BIBLIOGRAPHY

I. Manuscripts

Anleitung, Schiesspulver zu bereiten, Büchsen zu laden und zu beschiessen, 14.-15. Jhd. Codex germ. 600. München, Bayerische Staatsbibliothek.

Geschützabbildungen mit Beischriften von Johannes Formschneider aus Nürnberg und anderer, ca. 1460-1470. Codex germ. 734. München, Bayerische Staatsbibliothek.

Geschützbuch oder Zeugbuch Kaiser Maximilians, ca. 1500. Codex icon. 222. München, Bayerische Staatsbibliothek codex 141, Ambraser Sammlung, Wien, Österreichische Nationalbibliothek.

Description de l'artillerie de l'invincible empereur Charles Quint, 1552. Paris, Bibliothèque nationale.

MEYER, F., *Bichsenmeistery,* 1594. Cgm. 8143. München, Bayerische Staatsbibliothek.

Discours et dessins par lesquels s'acquiert la congnoissance de ce qui s'observe en France en la conduite et emploi de l'artillerie... par le capitaine VASSE-LIEU, *dit Nicolay Lionnais.* Début XVIIᵉ s. Ms. fr. 6994. Paris, Bibliothèque nationale.

GRIBEAUVAL, J. B. de, *Collection complète de la nouvelle artillerie construite dans les arcénaux de Metz et de Strasbourg,* Paris, Bibliothèque du Musée de l'armée.

GRIBEAUVAL, J. B. de, *Nouveau traité de construction d'artillerie,* 1780. Ms. fr. 9170, 9171. Paris, Bibliothèque nationale.

II. Monographs and specialised works

VALTURIUS, *De re militari.* 1472.

PFINTZING, M., *Theurdanck,* Nürnberg 1517.

TARTAGLIA, N., *Quesiti et Inventioni Diverse.* Venezia 1528.

FRONSPERGER, L., *Kriegsbuch,* mit Holzschnitten von Jost Amman. Frankfurt 1566, 1571, 1573.

COLLADO, K., *Practica manuale di artiglieria.* Venezia 1586.

MARCHI, F. de, *Pratica manuale dell'Artiglieria.* 1586.

BOURNE, W., *The Arte of Shooting in Great Ordnaunce.* London 1587.

CAPOBIANCO, A., *Corona e Palma militare di Artigliera.* Venezia 1598.

ERRARD, J., *La fortification démontrée et réduite en art.* Paris 1600.

RIVAULT, D., *Les Elemens de l'Artillerie.* Paris 1608.

UFANO, D., *Archeley, D.i. Gründl. und Eygenth. Bericht v. Geschütz und aller Zugehör.* Frankfurt 1615.

NORTON, R., *Of the Art of Great Artillery.* London 1624.

HONDIUS, H., *Description et brève déclaration de la fortification, de l'artillerie, des amunitions et vivres, des officiers et de leurs commissions, des approches, avec la manière de se défendre et des feux artificiels.* La Haye 1625.

The Gunner, Shewing the Whole Practice of Artillery transl. by Norton, R. London 1628.

Vraye Instructions de l'artillerie. Rouen 1628.

ROBERTS, J., *The compleat Cannoniere.* London 1639.

MALTHE, F. de, *Traité des feux artificiels pour la guerre et pour la récréation.* Paris 1640.

CERDA, T. de, *Lecciones de Artilleria.* Madrid 1644.

SIEMIENOWICZ, C., *Grand art d'artillerie.* Amsterdam 1651.

MORETTI, T., *Trattato dell'Artiglieria.* Brescia 1672.

BINNING, Th., *A Light to the Art of Gunnery.* London 1677.

MIETH, M., *Neue curieuse Beschreibung der gantzen Artillerie.* Dresden & Leipzig 1683, 1705, 1736.

GAYA, L. de, *Traité des Armes.* Paris 1678.

BUCHIERS, *Theoria et Praxis Artilleriae.* Nürnberg 1682.

MORETTI, T., *A General Treatise of Artillery,* transl. by J. Moore London 1683.

SURIREY de Saint-Rémy, *Mémoires d'artillerie,* 2 vol. Paris 1697.

BELIDOR, B. F. de, *Le bombardier français.* Paris 1731.

LEBATUT, *Traité contenant les proportions des canons.* Le Havre 1737.

MULLER, J., *Treatise of Artillery.* London 1757.

LEBLOND, *Artillerie raisonnée.* Paris 1761.

SCHEEL, de, *Mémoires d'artillerie contenant l'artillerie nouvelle ou les changements faits dans l'artillerie française en 1765.* Copenhague 1777.

DUTEIL, J., *De l'usage de l'artillerie nouvelle dans la guerre de campagne.* Metz 1778.

ROBINS, B., *Nouveaux principes d'artillerie,* commentés par Euler. Paris 1783.

MANSON, J. Ch. de, *Tables des constructions des principaux attirails de l'artillerie, proposées ou approuvées depuis 1764 jusqu'en 1789 par M. de Gribeauval,* 7 vol. Paris 1792.

MONGE, G., *Description de l'art de fabriquer les canons, faite en exécution de l'arrêté du comité de Salut public.* Paris An II (1794).

STEVENS, W., *A System for the Discipline of the Artillery of the United States of America.* New York 1797.

KOSCIUSKO, Th., *Maneuvers of Horse Artillery.* New York 1800.

LESPINASSE, *Essai sur l'organisation de l'arme de l'artillerie.* Paris an VIII (1800).

SMITH, A., *A Short Compendium of the Duty of Artillerists.* Worcester Mass. 1800.

A Treatise of Artillery containing a New System or the Alterations made in the French Artillery since 1765, trans. by J. Williams. Philadelphia 1800.

Aide-mémoire à l'usage des officiers d'artillerie de France, 2 vol. Paris 1801.

SCHARNHORST, G. von, *Handbuch der Artillerie,* 3 vol. Hannover 1804-1814.

— *Traité de l'artillerie,* 3 vol. Paris 1840-1843.

TOUSARD, L. de, *American Artillerist's Companion,* 3 vol. Philadelphia 1809-1813.

DARTEIN, Ch. M., *Traité élémentaire sur les procédés en usage dans les fonderies pour la fabrication des bouches à feu d'artillerie.* Strasbourg 1810.

ROUVROY, F. G., *Vorlesungen über die Artillerie,* 3 vol. Dresden 1811-1814.

A System of Artillery Discipline. Boston 1813.

LALLEMAND, H.D., *Traité d'artillerie.* La Nouvelle-Orléans ca. 1820.

COTTY, G. H., *Dictionnaire de l'artillerie.* Paris 1822.

POUMET, P. A., *Instruction sur l'artillerie de campagne.* Paris 1824.

A System of Exercise and Instruction of Field Artillery Including Maneuvers for Light or Horse Artillery. Boston 1829.

Règlement sur les manœuvres et les évolutions des batteries attelées. Approuvé par le Roi le 12 mars 1836. Paris et Strasbourg 1836.

BREITHAUPT, colonel, *Esquisse générale d'une nouvelle organisation de l'artillerie,* in: Spectateur militaire, tome XXIII, 1837.

MEYER, M., *Manuel historique de la technologie des armes à feu,* 2 vol. Paris 1837-1838.

KAMEKE, H. F., *Sammlung von Zeichnungen, die Einrichtung der materiellen Gegenstände der preussischen Artillerie darstellend, und Erläuterungen dazu,* 2 vol. Berlin 1837 & 1843.

MAZÉ, L. F., *Etat actuel de l'artillerie de campagne en Europe,* traduit de l'allemand, 7 vol. Paris 1838-1849.

HUGUENIN, U., *Description de la fabrication des bouches à feu en fonte de fer et des projectiles à la fonderie de Liège.* Paris 1839.

TIMMERMANS, C., *Essai d'un traité élémentaire d'artillerie,* 3 vol. Liège 1839.

Instruction for Field Artillery, Horse and Foot. Baltimore 1845.

Mazé, L. F., *Artillerie de campagne en France, description de l'organisation et du matériel de cette arme en 1845*. Paris 1845.

Espiard de Collonge, baron, *Artillerie pratique employée sous les règnes et dans les guerres de Louis XIV et Louis XV*, 2 vol. Paris 1846.

Instruction for Heavy Artillery. Washington 1851.

Instruction for Mountain Artillery. Washington 1851.

Coquilhat, cap, *Cours élémentaire sur la fabrication des bouches à feu en fonte et en bronze et des projcetiles d'après les procédés suivis à la fonderie de Liège*. Liège 1856.

Reports of experiments on the strength and other properties of metal for cannons; with a description of the machines for casting metal, and of the classification of cannons in services. Philadelphia 1856.

Berger, F., *Zeichnungen des Königl. Preuss. Artillerie Materials*, 2 vol. Berlin 1856-1857.

Anderson, R., *Evolutions of Field Batteries of Artillery*. New York 1860.

Gibbon, J., *The Artillerist's Manual*. New York 1860.

Holländisches Artilleriematerial. sl. 1861-1866.

Planches du matériel d'artillerie des Pays-Bas. sl. 1861-1879.

Andrews, R. S., *Mounted Artillery Drill*. Charleston 1863.

Instruction for Heavy Artillery. Washington 1863.

Roberts, J., *The Hand-Book of Artillery for the Service of the United States*, 5th ed. New York 1863.

Owen, C. H., *The Principle and Practice of Modern Artillery*. London 1871.

Beckerhinn, C., *Die Feld-Artillerie Österreichs, Deutschlands, Englands, Russlands, Italiens und Frankreichs in Bezug auf ihre Bewaffnung, Ausrüstung, Organisation und Leistungsfähigkeit*. Wien 1879.

Atlante del materiale di Artiglieria adottato dal Ministero della guerra nell' anno 1884. Roma 1884.

Dredge, J., *Modern French Artillery*. New York 1892.

Lloyd, E. W., & Haddock, A. G., *Artillery: Its Progress and Present Position*. Portsmouth 1893.

Pratt, S. C., *Field Artillery*. London 1895.

Handbook of the 13-Pdr. Q. F. Gun Land Service. London 1913 & 1914.

Handbook for 15-Pdr. B. L. Gun, Marks II to IV and Carriages, Marks II and IV, and Wagon, and Limber, Mark IV. London 1914.

Hime, H. W. L., *The Origin of Artillery*. London 1915.

Handbook of the 4.5-in. Q.F. Howitzer Land Service. London 1916.

Gascouin, F. E., *L'évolution de l'artillerie pendant la guerre*. Paris 1920.

Thouvenin, L., *L'artillerie nouvelle, munitions, tir, matériels*. Paris 1921.

Rathgen, B., *Das Geschütz im Mittelalter*. Berlin 1928.

The War Office Handbook for the Q.F. 3.7-in. Mark I Howitzer or Marks I, II, & III Carriages Land Service. London 1930.

Culmann, F., *Tactique d'artillerie. Matériels d'aujourd'hui et de demain*. Paris 1937.

Jakobsson, Th., *Artilleriet under Karl XII*. Stockholm 1943.

Egg, E., *Der Tiroler Geschützguss 1400-1600*. Tiroler Wirtschaftsstudien N° 9. Innsbruck 1961.

Lombarès, M. de, *Un certain Konrad Hausser, inventeur du canon de 75*, in: Revue historique de l'armée, 2/1969.

III. General works on artillery

Graewenitz, W. von, *Organisation und Taktik der Artillerie*. Berlin 1824.

— *Traité de l'organisation et de la tactique de l'artillerie et histoire de cette arme*. Paris 1831.

Brunet, J. B., *Histoire générale de l'artillerie*, 2 vol. Paris 1842.

Napoléon III & Favé, I., *Etudes sur le passé et l'avenir de l'artillerie*, 6 vol. Paris 1846-1871.

Lorédan, L. *Origine de l'artillerie*, planches autographiées d'après les monuments des XIV et XV siècles, 2 vol. Paris 1863.

Tennet, J. E., *The Story of Guns*. London 1864.

Angelucci, A., *Documenti inediti per la storia delle armi da fuoco italiane*. Turin 1869.

Essenwein, A., *Quellen zur Geschichte der Feuerwaffen*, 2 vol. Leipzig 1877.

Boeheim, W., *Handbuch der Waffenkunde*. Leipzig 1890.

Campana, J., *L'Artillerie de campagne 1892-1901*. Paris 1901.

Jocelyn, J. R. J., *The History of the Royal Artillery*. London 1911.

Gohlke, W., *Geschichte der gesamten Feuerwaffen bis 1850*. Sammlung Goeschen. Leipzig 1911.

Herr, F. G., *L'artillerie: ce qu'elle a été, ce qu'elle est, ce qu'elle doit être*. Paris 1923.

Bruchmüller, G., *Die Artillerie beim Angriff im Stellungskrieg*. Charlottenburg 1926.

Graham, C. A. L., *The Story of the Royal Regiment of Artillery*. London 1928.

Calwell, Ch., *History of the Royal Artillery 1860-1914*, 2 vol. London 1931 & 1937.

Masset, M.A., *Les grands maîtres de l'artillerie*. Paris 1933.

Montu, C., *Storia dell'artiglieria italiana*, 2 vol. 1934-1935.

Challéat, J. *Histoire technique de l'artillerie de terre en France pendant un siècle (1816-1919)*. Paris 1935.

Menu, général, *L'artillerie dans la bataille*. Clermont-Ferrand 1939.

Wilson, A. W., *The Story of the Gun*. Woolwich 1944.

Druène, lt-colonel, *Deux siècles d'histoire de l'artillerie française*, in Revue historique de l'armée, 2/1954.

Tunis, E., *Weapons*. New York 1954.

Carman, W. Y., *A History of Freimans*. London 1955.

Downey, F., *Sound of the Guns*. New York 1955.

Hogg, O. F. G., *English Artillery 1326-1716*. London 1963.

Pope, D., *Guns*. London 1965.

Müller, H., *Deutsche Bronzegeschützrohre*. Berlin 1968.

Peterson, H. L., *Round Shot and Rammers*. Harrisburg 1969.

IV: General works

Malthus, F., *Pratique de la guerre*. Paris 1646.

Diderot, D., *Encyclopédie ou Dictionnaire des Sciences, des Arts et des Métiers*. Paris 1751-1777.

Guibert, J. A., *Essai général de tactique*, 2 vol. London 1772.

Smith, G., *Universal Military Dictionary*. London 1779.

Jomini, A. H., *Histoire critique et militaire des guerres de la Révolution, de 1792 à 1803*, 15 vol. Paris 1814-1824.

O'Connor, J. M., *A Treatise on the Science of War and Fortification*. New York 1817.

Martini, G., *Architecture civile et militaire*. Turin 1841.

Todleben, E. de, *Défense de Sébastopol*, 4 vol. St. Petersbourg 1863-74.

— *Die Vertheidigung von Sebastopol*. Übersetzung aus dem Russischen, 4 vol. St. Petersburg 1864-1872.

Fortescue, J. W., *History of the British Army*, 13 vol. New York and London 1899-1930.

Foch, général F., *Des principes de la guerre*. Paris 1911.

Schwarte, M., *Die militärischen Lehren des grossen Krieges*. Berlin 1920.

Heigl, major, *Militärwissenschaftliche Mitteilungen*. Österreich 1930.

Quinton, R., *Maximes sur la guerre*. Paris 1930.

Field, C., *Echoes of Old Wars (1513-1854)*. London 1934.

Rommel, E., *Infanterie greift an*. Potsdam 1937.

Fuller, J. F. C., *Armaments and History*. London 1946.

Guderian, H., *Erinnerungen eines Soldaten*. Heidelberg 1951.

— *Souvenirs d'un soldat*. Paris 1954.

Falls, C., *A Hundred Years of War*. London 1953.

Klass, G. von, *Krupps*. London 1954.

Millis, W., *Armies and Men*. London 1958.

Montgomery of Alamein. *A History of Warfare*, London 1958.

Howard, M., *The Franco-Prussian War*. London 1961.

Scott, J. D., *Vickers. A History*. London 1962.

Gibbs, P., *The Battle of Alma*. London 1963.

Falls, C., *Great Military Battles*. London 1964.

Batty, P., *The House of Krupp*. London 1966.

ACKNOWLEDGEMENTS

The Publishers are grateful to all the curators of museums, librarians, the directors and administrators of official bodies and private societies who have helped them to assemble the information and the illustrations for this work. In particular, they would like to thank Colonel EMG Daniel Reichel, Artillery Instructor and Director of the Federal Military Library at Berne, and Major R. L. Ellis FRSA, RA (Rtd) and his family for all their enthusiastic help and advice.

ILLUSTRATION SOURCES

Agence Tass, 198 right; Air Force, Space digest, 197; Bayr. Staatsbibliothek, Handschriftabteilung, München, 11, 14, 16, 17, 20, 24, 25; Bayr. Staatsgemäldesammlungen, München, 42, 43, 45; v. Bergen, 169 centre, 170 bottom, 176, 177, 179; Bern. Historisches Museum, Bern, 73; Biblioteca Ambrosiana, Milano, codex atlanticus, 29; Bibliothèque cantonale et universitaire, Lausanne, 75 top right, 92-95, 155 bottom, 163; Bibliothèque militaire fédérale, Berne, 87 bottom, 130, 131, 133, 135, 136, 143, 150, 151 top, 151 left, 154, 156-158, 160-162, 166, 168 top left, 172, 173, 175 top right, 181, 183, 184 top, 193 top and centre, 194, 195; Bibliothèque du Ministère de la Guerre, Paris, 53, 67, 150; Bibliothèque du Musée de l'Armée, Paris, 88, 89; Bibliothèque nationale, Paris, 55-59, 63, 71, 84, 85, 87 top, 99, 129, 140, 148, 169 bottom, 170 top, 186, 187, 188 top, 190 top, 191 top and left; British Museum, 13; Bofors, Sweden, 193 bottom; Bundesministerium der Verteidigung, Bonn, 198 left; Château de Chantilly, 33; Garzanti, Milano, 168 top right; Germanisches Nationalmuseum, 46, 66, 68, 69, 74-79, 81-83; Graphische Sammlung Albertina, Wien, 50; Heeresgeschichtliches Museum, Wien, 141; Imperial War Museum, London, 159; Mansell Collection, 121, 123; Musée de l'Armée, 117; Musée d'Arras, 155 top; Musée Atger, Montpellier, 70; Musée de la Guerre, Paris-Vincennes, 167, 168 bottom, 169 top, 174, 175 top left, 184 bottom, 185; Musée du Louvre, Paris, 61; Musée de Versailles, 98, 102, 103, 106, 111, 112; Museo Storico Navale, Venezia, 49, 90; The Museum of Modern Art, New York, 147, 148; Österreichische Nationalbibliothek, Wien, 21; Royal Artillery, Bramcote, 129 top; Science Museum, London, 145; Staatsbibliothek, Berlin, Bildarchiv (Handke), 149, 153, 171; Staatliche Graph. Sammlung, München, 19, 39-41, 47; Stadt- und Universitätsbibliothek, Bern, 72, 75 top left, 86; Ullstein, Berlin, 145 bottom; Universitätsbibliothek, Basel, 12, 15, 18; USIS, Paris, 188 bottom, 189; Zentralbibliothek, Luzern, 28.

PHOTOGRAPHIC CREDITS

Agence Tass, 198 right; Air Force, Space Digest, 197; Archives photographiques, Paris, 70; Bayr. Staatsbibliothek, München, 11, 14, 16, 17, 20, 24, 25; Bayr. Staatsgemäldesammlungen, München, 42, 43, 45; v. Bergen, 169 centre, 170 bottom, 176, 177, 179; Bernisches Historisches Museum, Bern, 73; Biblioteca Ambrosiana, Milano, 29; Bulloz, Paris, 106 bottom; Bundesministerium der Verteidigung, Bonn, 198 left; Cie des Arts photomécaniques, Paris, 102, 103; Courtauld Institute of Art, London, 13; La Documentation française, 169 bottom, 170 top, 186, 187, 188 top, 190 top; Edita, Lausanne, 12, 15, 18, 53, 55-59, 63, 67, 71, 72, 75 top, 84-89, 92-95, 99, 129-131, 133-136, 140, 141, 143, 150-151, 154, 155 bottom, 156-158, 160-163, 166, 168 top left, 172, 173, 175 right, 181-183, 184 top, 190 bottom, 191, 192, 194-195, 196; Garzanti, Milano, 168 top left; Germanisches Nationalmuseum, Nürnberg, 46, 66, 68, 69, 74, 75 bottom, 76-79, 81-83; Giraudon, Paris, 33, 106 top; Graphische Sammlung Albertina, Wien, 50; Imperial War Museum, London, 159; Mansell Collection, London, 121, 123; Musée de l'Armée, Paris, 117; Musée d'Arras, 155; Musées Nationaux, Paris, 61, 98, 111, 112; Museo Storico Navale, Venezia, 49, 96; The Museum of Modern Art, New York, 147, 148; Musée de la Guerre, Paris-Vincennes, 167, 168 bottom, 169 top, 174, 175 left, 184 bottom, 185; Österreichische Nationalbibliotek, Wien, 21; Science Museum, London, 145; Staatsbibliotek, Berlin, 149, 153, 171; Staatl. Graphische Sammlung, München, 19, 39, 40, 41, 47; Ullstein, Berlin, 145; USIS, Paris, 188 bottom, 189; Zentralbibliotek, Luzern, 28.

This book was printed by
Imprimeries Réunies S.A., Lausanne,
and bound by Maurice Busenhart, Lausanne

Printed in Switzerland